Discovering the *Living Ocean*

LABORATORY AND FIELD ACTIVITIES IN MARINE BIOLOGY

Paul Detwiler
Mesa College of San Diego

Kendall Hunt
publishing company

Always pay careful attention to safety procedures when performing any laboratory or field activity. To reduce your risk of injury, explicitly follow all instructions as written in this manual and as provided by your instructor. While safety precautions have been considered in the writing and preparation of these experiments, these activities are conducted at your own risk. The publisher and author cannot be held liable for any loss or damage incurred resulting from or related in any way to the execution of the procedures described herein, or to the activities themselves, regardless of the form of action.

Kendall Hunt
publishing company

www.kendallhunt.com
Send all inquiries to:
4050 Westmark Drive
Dubuque, IA 52004-1840

ISBN 978-0-7575-8485-5

Printed in the United States of America
10 9 8 7 6 5 4 3

TABLE OF CONTENTS

FIELD STUDIES

APPENDICES

PREFACE:

I wrote *Discovering the Living Ocean* to provide a suitable manual in marine biology that is engaging, easy to read, and relevant for nonscience majors in community college as well as high school students. My approach was to prepare activities that challenge and inspire students to learn more about the Ocean that all of us share and depend on in so many ways.

This manual is appropriate for students who have no previous background in biology. The variety of exercises and activities allow instructors to select ones that are feasible for their classroom use, given their students' abilities, the time constraints of the class period, and the types of resources available in the classroom and in the field. I have included activities that expose students to some of the multifaceted disciplines that marine biologists must have a working knowledge of: microbiology, zoology, phycology, anatomy, physiology, ecology, and oceanography (to name a few).

The *Teacher's Guide* includes lists of supplies, organisms, and equipment needed for each activity (and lists the sources of those materials); provides lab safety considerations; offers suggestions for conducting the lab and/or field procedures, and recommends relevant websites and video resources. The *Teacher's DVD* contains additional video clips that may be used in assessment.

To the Student:

Growing up on the California coast, I don't remember at what point I became interested in the Ocean and its life forms. I recall certain vivid childhood memories: a stingray rocketing through the shallows in the surf zone towards my feet; the rubbery smoothness of the flank of a captive bottlenose dolphin; the burning sting from a beached jellyfish. Marine biology has always been exciting and inspirational to me, and I've been fortunate as an educator to have the opportunity to help students discover their own sense of enthusiasm and interest for the Ocean and its varied, amazingly-adapted inhabitants.

To study marine biology means that you have to take a "big picture" view of our planet, since marine science is an *integrative* discipline: you must understand elements, ideas, and concepts from different fields of science and synthesize them to help you understand how and why the life forms in the oceans are the way they are—how they live, how they adapt to their environment, and how they interact with each other. Thus, you need to understand a little bit of astronomy, meteorology, oceanography, chemistry, physics, zoology, botany, and ecology, and develop a *holistic* view of how the Ocean maintains life on our planet. Studying marine biology also means developing focused observational skills when examining marine organisms and their environments. The laboratory and field activities in this manual will guide you in making close, detailed observations that lead to questions about what you are seeing, and then to hypotheses that you develop and possibly test to help answer those questions. Approach each activity in this manual with these two scales of observation in mind.

Here are a few suggestions to help you be successful in your marine biology class.

- There is an accompanying DVD to this manual. If you have it, watch the relevant video chapters before each lab occurs, so that you understand the procedures being demonstrated. The more prepared you are going into the lab or field, the more efficient you will work, and the more you will accomplish in the limited time available during the class period.
- Use the glossary to learn any terms that you are unfamiliar with. **Underlined** and **bold** words are important to learn, and are either defined in context or in the glossary.
- Your manual is a workbook. In it, write your answers to all the questions posed in each activity. When you answer the questions, you are creating your own study guide to help you remember the main ideas and conclusions of the lab or field exercise. If you don't

attempt to answer the questions, it is likely that you will not do well on a later quiz or test based on the lab or field activity. Some questions involve discussion with fellow students, while others may involve outside reading.
- Follow all safety guidelines provided by your instructor, and those demonstrated in the DVD and given in the manual.

Knowing about life in the Ocean is vitally important for everyone, not just for those who make their living from it or those who live on the coastline. Everyone breathes oxygen produced by oceanic life forms, and the health of our planet depends on a healthy Ocean. But now the Ocean is in crisis and immediate action is needed to prevent it from being irreparably damaged. It would be a tragedy if yours is the last generation to have a chance of viewing the wonders of a healthy coral reef or to share the Ocean with sharks, whales, sea turtles, and large fish. If all of us as a society do not work to reverse present trends, these unfortunate scenarios may well come true. The marine biologists emerging from your peer group will be challenged (even more so than the biologists of today) to use scientific knowledge to find new solutions for the environmental problems humans have inflicted on the Ocean, and to find ways to protect and conserve oceanic life so that future generations can enjoy a diverse, awe-inspiring undersea world as previous generations have.

I hope that your appreciation, curiosity, and respect for the biological world around us will be heightened by what you learn in this manual, and that you will be inspired to do your part to protect the oceans and the world of nature, far into the future.

Traditional Polynesian seafarers from the numerous, remote islands scattered throughout the Pacific had a saying:

The sea does not separate us, it unites us.

We as one nation among many nations in the global community would do well to remember this sentiment to foster international collaborations to manage, preserve, and protect the Ocean.

–P.D.

Acknowledgements:

I was encouraged throughout the writing process by many people who provided unflagging help while I brought this manual to life. Special thanks to Sean, Geoff, Michael, Joe, James and Lorin for their friendship, humor, and continued support to keep the vision on track.

I am grateful to my publishers for their cheerful guidance and for the opportunity to foster awareness, curiosity, and appreciation for the Ocean in a new generation of marine biology students.

Gratitude is owed to M. Tiffany, B. Gladden, D. Dexter, R. Vaca, M. Russell, C. Bauder, M. Van Patten, J. Detwiler, M. Chen, and D. Rudie for use of their photographs, specimens, video footage, equipment, and technical expertise. M. Kwon provided several illustrations. Unless otherwise indicated, all photography and videography is the author's.

The following reviewers generously contributed their comments and suggestions to improve the manual. Thanks are due to:

R. Alquezar, University of Central Queensland, Gladstone, Australia; S. Bartram, San Diego Supercomputer Center; J. Connor, Montery Bay Aquarium Research Institute; D. Dexter, San Diego State University; N. Dow, Mosley HS; T. Ebert, Oregon State University; T. Fish, National Park Service; C. Fowler, Garfield HS; M. Friedman, Animo Leadership Charter HS; H. Gastil, San Ysidro HS; L. Harrington, James Cook University, Townsville, Australia; B. Hines, Mountain Brook HS; M. Jones, Monmouth Academy; S. Martin, University of Maryland, College Park; S. Moore, Croatan HS; J. Newsome, San Diego State University; K. Reifel, University of Southern California; N. Scola, New England Aquarium; B. Swan, University of California, Santa Barbara; A. Wasser, American Heritage Marine Institute.

SAFETY PRECAUTIONS FOR SCIENCE ACTIVITIES

To the student: You must read and remember the following safety precautions listed here and at the start of any activities in the manual. Use good judgment and pay attention to all steps of the procedures to reduce the risk of accidents and/or injury. If you have questions about a potentially dangerous material or procedure, stop what you are doing and ask your teacher for further directions.

Preparation (Before the Activity)

- Be aware of hazards and know the locations of the exits and the first aid kit, and know how to use the fire blanket, fire extinguisher, and the eyewash station. Use any of the additional safety equipment that is provided to you.
- Read the instructions for the entire experiment and all procedures before you begin any work. If you have questions, ask your teacher.
- Before you begin, all surfaces of the workspace must be clean and cleared of personal items, backpacks, books, etc.
- Always obtain your teacher's permission before you begin the activity.
- Prepare yourself:
  - Do not wear loose clothing or open-toed shoes in lab.
  - Long, baggy shirt sleeves should be rolled up, or remove the item of clothing.
  - Large jewelry (rings, dangling bracelets) should be removed.
  - Tie back long hair.
  - Wear safety glasses (or goggles) and a laboratory apron.

During the Activity

- Follow all the directions of a procedure in order, one step at a time.
- Keep your workspace neat, clean, and dry.
- If chemicals are spilled, notify your teacher and clean up the spill as directed.
- Never eat or drink anything in the laboratory.
- Never taste any substance unless your teacher tells you explicitly that it is safe to do so.

Equipment Usage

- Set up your apparatus in the center of the workspace, away from the edge of the tabletop.
- Check glassware for cracks and sharp edges before using.
- Use caution when handling sharp or pointed instruments.
- When inserting and removing devices from electrical outlets, pull their plugs, not their cords.
- Never pipette liquids by mouth, always use a suction device.
- Notify your instructor if glassware breaks. Sweep it up immediately and dispose of broken glass in the appropriate container.
- Clean glassware after use, as your instructor advises.

Chemical Safety

- Wear safety glasses, a laboratory apron, and gloves.
- Read all labels carefully before mixing chemicals.
- Never taste any substance or inhale chemical vapors.
- Avoid chemical contact with your eyes and skin. Do not rub your eyes or touch your face when handling chemicals.
- If foreign matter gets in your eyes or on your skin, flush with water from the sink or at the eyewash station.

- If you spill any chemical, report it immediately to your teacher.
- Wash your hands before and after using chemicals.
- Report any injury or accident, no matter how trivial it seems, to your instructor.

When Heating Materials by Flame (or with an Electric Hot Plate)

- Know the location of the fire extinguisher, and how to operate it.
- Use gloves, safety goggles, an apron, safety tongs, and heat-resistant mittens when heating materials.
- Keep loose clothing and other flammable objects (papers, notebooks) away from the heat source.
- Keep your face away from heated glassware.
- When heating a test tube of boiling liquid, always point its mouth away from yourself and other people.
- Continuously monitor what is being heated. Never leave a sample unattended while it is being heated.
- Only take measurements on samples after they have completely cooled.
- If you burn yourself, flood the burned area with cold water, and tell your teacher immediately.
- If your clothing catches fire, smother it with the fire blanket or a coat, or get to the sink or eyewash station and flood it with water. **Do not run**, as this will fan the flames.
- Turn off hot plates, extinguish the Bunsen burners, and turn off gas lines when you have finished.

Cleaning Up After the Activity

- Turn off running water, and shut off gas lines. Turn off and unplug electrical devices.
- Dispose of chemicals and other materials as your instructor advises. Return unused chemicals to their proper containers.
- Don't dispose of any materials down the sink unless your teacher instructs you to do so.
- Thoroughly clean the work area and any glassware that you used.
- Wash your hands after working in the laboratory.
- Wait for your instructor's permission to be dismissed from the room.

Safety Contract

I, ____ , have read and understand the safety rules and first aid information as described above. I will observe all of these safety rules at all times when engaging in science activities in the field and classroom.

____ ____________________

Signature *Date*

LABORATORY 2

Examining the Properties of Seawater

INTRODUCTION:

Understanding the physical nature of the marine environment is a fundamental component toward understanding the biology and ecology of marine organisms, because all species are affected in some way by the physical conditions of their environment. **Abiotic** factors, such as temperature and availability of water, are known to limit the abundance and distribution of species in terrestrial environments. For example, most types of mosses grow abundantly in moist, cool areas, but not in hot, dry areas. Similarly, certain factors in the ocean environment, such as seawater temperature and its chemical composition, are extremely important in influencing what kinds of marine organisms are found in certain places within the oceanic realm, and how abundant those organisms may become. Thus, when marine biologists study organisms in the field, they also collect data on these key environmental variables: seawater temperature, salinity, and the amount of oxygen dissolved in the seawater.

This laboratory will familiarize you with several techniques commonly used by marine scientists to describe a few basic physical and chemical characteristics of seawater. After completing this laboratory exercise, you will be able to do the following:

- Measure salinity and dissolved oxygen content of water samples using several different methods.
- State the relationships between temperature, salinity, and density.
- Describe how seawater density affects buoyancy.
- Explain how temperature and salinity affect dissolved oxygen content in seawater.
- Use graphs to analyze density and oxygen solubility.

EXERCISE 1: TEMPERATURE AND DENSITY

Marine organisms live surrounded by water, and so they are strongly influenced by its properties. One property, water **density**, is of key importance to life forms in the oceans. Density is a property of every kind of material, and is expressed by the formula:

$$\text{Density} = \frac{\text{Mass}}{\text{Volume}}, \quad \text{where:}$$

Mass = Weight of material or object (in grams, g)

Volume = Amount of space that the material occupies (in cubic centimeters, cm^3)

For example, calcium carbonate ($CaCO_3$), a major structural material in coral, seashells, and the bones of marine vertebrate animals, has a density of 2.83 g/cm^3. Thus, a cube of $CaCO_3$ that is 1 cm in length, width, and height will weigh 2.83 grams. Recall that 1 cm^3 is equivalent in volume to 1 ml (see Appendix A, Metric System). By definition, the density of pure water is 1.000 g/ml at 4° Celsius, meaning that 1 ml of pure water weighs 1.000 g at that standard temperature. Temperature is specified because water density can be less or greater than 1.000 g/ml, depending on the temperature. In this activity, you will seek to answer the question, "How does temperature affect water density?"

Write your hypothesis here:

Procedure:

1. Wrap a penny loosely with a small square of aluminum foil.
2. Fill a 1000-ml graduated cylinder with ice water to the 1000 ml mark. Make sure there are no pieces of ice in the cylinder.
3. Use a stopwatch to record how long the penny takes to sink to the bottom of the cylinder. Hold the penny just at the water surface and begin timing immediately when you release it. Stop the clock when any part of the wrapped penny touches the bottom. Record time to the nearest 0.1 second, and enter your result in the following table.
4. Recover the penny with the wire retrieval wand, and repeat Step #3 for a total of 10 trials.
5. Using the same penny, repeat the experiment using a cylinder filled with 50°C water, and enter your results below.

Table 1. Comparison of Sinking Rates of a Negatively Buoyant Object (A Penny).

Trial #:	5°C water time (sec)	50°C water time (sec)
1		
2		
3		
4		
5		
6		
7		
8		
9		
10		
Average time:		
Difference (in sec) between average sinking times:		

Conclusions/Interpretation:

How is the penny's sinking time affected when water temperature is raised?

In which liquid was the penny's sinking speed greater?

Which type of material would it be easier and faster for a **nektonic** marine organism to swim through, a higher- or lower-density liquid?

Which water temperature is the lower-density liquid?

Therefore, does an increase in temperature *increase* or *decrease* water density?

EXERCISE 2: SALINITY AND DENSITY

Seawater is not pure water; many substances are dissolved in it that contribute to its overall **salinity**, or the total amount of dissolved salts that the water contains. The salinity of ocean water is affected by a variety of factors, including precipitation (as rain and snow), evaporation, drought conditions, and the proximity of freshwater inflows, such as the mouths of rivers. In this activity, you will seek to answer the question, "How does salinity affect water density?"
Write your hypothesis here:

Procedure:

1. Using a balance, weigh an empty 100-ml volumetric flask to the nearest 0.001 g, and record its weight in Column A of the table below.
2. Remove the flask and carefully fill it to the neck line with room temperature distilled water, then reweigh it. To accurately measure liquid volumes, the bottom of the **meniscus** should touch the measure mark, as in Figure 1. Record the weight in Column B.
3. Repeat this procedure with a second flask: first weigh it empty, then after it has been filled to the 100-ml mark with room temperature seawater.
4. Determine the weight of water added by subtracting the value of Column A from Column B. Record this value in Column C.
5. Calculate the density of both water samples, as Column D indicates.

Figure 1. Accurate measurements of volume are taken when the bottom of the meniscus (shown by arrow) touches the graduated mark on the flask's neck.

Table 2. Comparison of Density between Water Samples of Varying Salinity as Determined by Weight.

	A	B	C	D
	Weight of flask (g)	Weight of flask + 100 ml water (g)	Weight of water (g) (= B – A)	Water Density (g/ml) (= C / 100 ml)
Distilled water				
Seawater				
Mixture				

6. Now repeat the experiment, but use a mixture of 50 ml seawater and 50 ml of distilled water in the volumetric flask. What is your prediction for the new density?

Conclusions/Interpretation:

Which water sample is denser? Why?

Did the results you obtained support your hypothesis or not?

EXERCISE 3: DETERMINING SALINITY

Investigation: How much salt is in seawater? Estimate how many level teaspoons of salt (sodium chloride) are needed to make a saline solution that approximately matches the salt concentration of ocean water.

Procedure:

1. Measure 500 ml of distilled water or drinking water into a beaker.
2. Add 1 level half-teaspoon of sodium chloride (stirring to dissolve after each addition).
3. Taste the saltwater solution and add more salt as needed, 1 level half-teaspoon at a time.
4. When you decide that the saline solution tastes like seawater, record your result here: ________ tsp.

linity is often expressed in parts per thousand (ppt). The approximate salinity of open ocean er is 35 ppt, meaning that there are 35 parts of salt dissolved in 1000 parts of water. If you h. 1000 ml of water, how many ml of sodium chloride must you add to make a saltwater solutio. hat matches the salinity of ocean water? ____________ ml.
Hov any teaspoons make this amount? Use a graduated cylinder to find out.

Record your result here: ________ tsp.

Does thi: easured value agree with the salinity level you estimated by taste?

Next, you w. etermine the actual amount of salts dissolved in a sample of seawater.

Procedure:

1. etermine the weight of an empty 100-ml beaker. Record this amount in Column A of Table 3 below.
2. Measure 20 ml of seawater using a graduated cylinder, and pour it into the 100-ml beaker. Weigh the beaker again, and record this new value in Column B.
3. Heat the beaker on a hot plate or over a burner until nearly all the water has evaporated; then loosely cover the beaker with aluminum foil to prevent salt loss from sputtering and bubbling as the last of the water boils off.
4. Remove the beaker from the heat source with tongs, remove the foil cover, and let the beaker cool completely. When cool, weigh it again, and record the data in Column D.
5. Calculate the weight of the seawater, as Column C indicates.
6. In column E, calculate the amount of salts contained in the 20-ml seawater sample by subtracting Column A from Column D.
7. To calculate the percent salt in the seawater sample, divide Column E by Column C and multiply the result by 100.

Table 3. Determination of Salinity by Evaporation.

A	B	C	D	E	F
Mass of empty beaker (g)	Mass of beaker and water (g)	Mass of water (g) (= B – A)	Mass of beaker and salt residue (g)	Mass of salts (g) (= D – A)	% of salt = (E / C) x 100

What is the salinity of your water sample (in percent and in parts per thousand)?

Suggest two sources of error in this procedure that would cause students using the same water sample to end up with different values for salinity.

Other Methods of Determining Salinity:

Hydrometer: A hydrometer is a blown-glass tube of a particular weight and is designed to float at a certain level in a water sample, which corresponds to the sample's density.

Refractometer: A refractometer is an optical device that measures the degree of light refraction (the bending of light rays) as light passes through a water sample. The refractive index of the water sample is adjusted for temperature. Salinity is either read directly in the instrument, or is derived from a table of refractive index values.

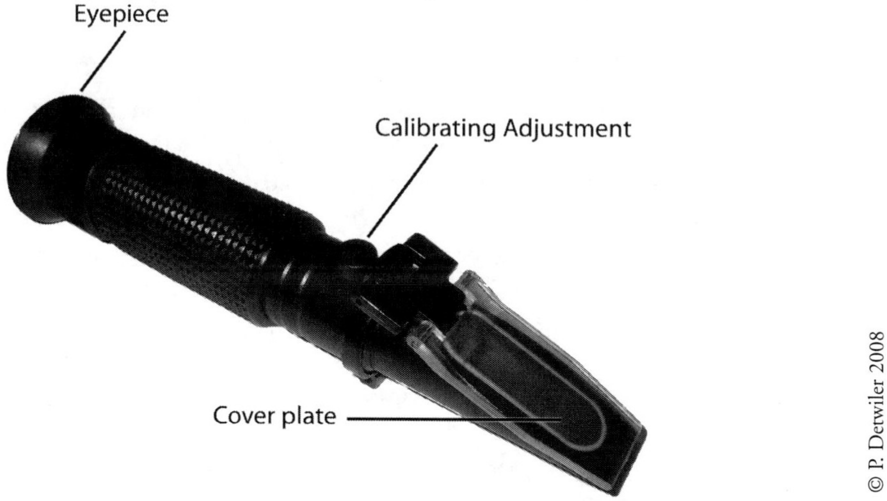

Figure 2. The refractometer measures salinity in a water drop placed under the cover plate. When viewed through the eyepiece, a line appears on an interior scale indicating the sample's salinity.

Salinometer: In practice, oceanographers measure salinity by electrical conductivity. An electric current is passed through the seawater sample, and the resistance of the water sample to the current is recorded. Resistance is directly proportional to the concentration of dissolved salts in the sample.

If you have access to any of these tools for measuring salinity, your instructor will demonstrate their operation.

EXERCISE 4: SALINITY AND BUOYANCY

In this activity, you will investigate the effects of salinity on the buoyancy of a submerged object. Marine organisms live surrounded by seawater of varying density. If an organism is *less* dense than the surrounding seawater, what might happen to that organism?

Conversely, if an organism's density is *greater* than that of the surrounding seawater, what will happen to the organism?

Furthermore, if an organism's density is *equal* to that of seawater, the organism will maintain its vertical position in the water column, remaining suspended at a particular depth. For the previous examples, marine scientists would describe the organisms as being **positively buoyant**, **negatively buoyant**, and **neutrally buoyant**, respectively.

Procedure:

1. Fill a 500-ml beaker 3/4 full with tap water. Use the graduated cylinder to measure the volume of water you added, and record this volume in Table 4, below.
2. Gently lower a chicken egg into the water with the spoon. Does it float? Record your observations in Table 4.
3. Remove the egg from the beaker. Measure 4 ml of sodium chloride using a small graduated cylinder, and add it to the beaker. Stir to dissolve.
4. Gently place the egg into the beaker as before, and record your observations.
5. Repeat the salt additions in 4-ml increments until the egg becomes neutrally buoyant.

Table 4. Effect of Salinity on the Buoyancy of an Egg.

Temperature of water: ____________ °C
Volume of water in beaker: ____________ ml

Total ml of salt added:	Buoyancy of Egg/Results Seen:
0	
4	
8	
12	
16	
20	
24	
28	

Conclusions/Interpretation:

You have seen how the egg's buoyancy changed during the experiment. Did the density of the egg change during the experiment?

Does increasing the amount of salt *increase* or *decrease* the density of water?

How can you tell that the water density is increasing?

How much salt was necessary for the egg to become neutrally buoyant?

Calculate the salinity of the water when the egg became buoyant.

Now compute how much salt would be necessary to float your egg in 1000 ml of water. How does this value compare to the salinity of seawater (greater/lesser/the same)? Would a chicken egg be negatively, neutrally, or positively buoyant in seawater?

EXERCISE 5: INFLUENCE OF TEMPERATURE ON SALINITY AND DENSITY

The salinity of seawater decreases near river mouths, as ocean water is diluted by the freshwater input. Seawater salinity can increase within bays and shallow enclosed water bodies, where evaporation is high and rainfall or other freshwater inflows are low, and where sea ice forms. These salinity changes affect water density, and sometimes the combination of salinity and temperature results in an additive effect: both variables act together to increase water density. At other times, these variables each exert their influence in opposite ways, lowering density. Fortunately, the relationships between temperature, salinity, and density have been determined, and these are shown in a T-S-D diagram (Figure 3). In this activity, you will use the diagram to explore the relationships between temperature and salinity as they jointly affect water density. You can determine any one of these 3 factors if the other 2 factors are known for a water sample.

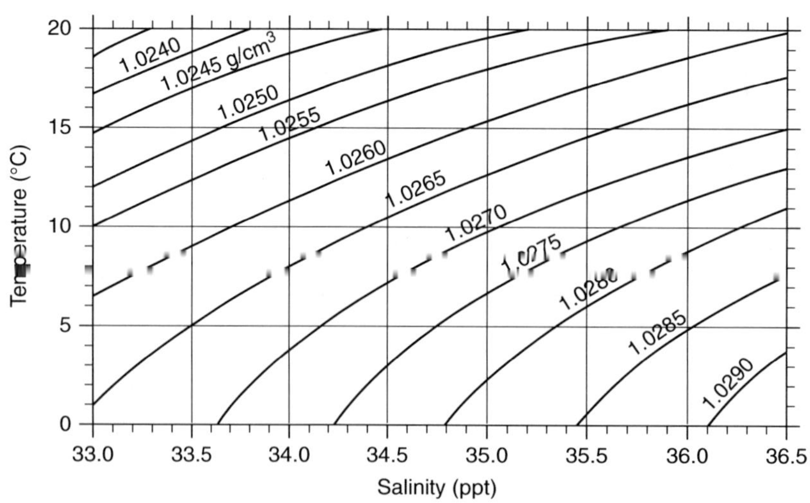

Figure 3. Relationships between Temperature, Salinity, and Water Density. The contour lines represent lines of equal density, reported as g/cm^3.

How do you use the diagram to determine the density of a water sample, given its temperature and salinity?
Example: A seawater sample has a temperature of 11.5°C and salinity of 34.0. Determine its density.
Solution: Follow the two data points along their x and y axes to the spot where the coordinates intersect. In this case, they meet on the line of constant density 1.026; therefore, density of this water sample is 1.026 g/cm^3.

Example: A seawater sample has a density of 1.0270 g/cm^3 at 10°C. Determine its salinity.
Solution: In this case, extend an imaginary line from 10° on the y-axis value across the diagram until it intersects the 1.0270 contour line. Trace the line down to the x-axis, and you see the corresponding salinity value is 35 ppt.

Practice exercises using Figure 3:
Make a dot on the TSD diagram corresponding to a seawater sample with a temperature of 8°C and a salinity of 34.0 ppt. Label this point with the letter A.

1. What is the density of this sample? ___________ g/cm^3.
   Now imagine that this seawater slowly heats up to 18°C, but the salinity remains unchanged.
2. What is its new density? ___________ g/cm^3.
   Now imagine that the temperature of sample A remains constant at 8°C, but its salinity increases from 34.0 to 35.9.
3. What is its new density? ___________ g/cm^3.
   Mark two more dots, one at 20°C and 34.5 ppt (label this B), and the other at 10°C and 34.5 (label it C).
4. What is the density of B? ___________ g/cm^3 of C? ___________ g/cm^3
   Now what would be the density of a new water sample composed of equal volumes of samples B and C mixed together? Mark a triangle on the diagram to reflect the new value and label it D.
5. What is this new density? ___________ g/cm^3
6. Rank samples B, C, and D in increasing order, from lowest to highest density:

Density differences are responsible for worldwide circulation patterns of seawater below the ocean surface, and for the vertical mixing of surface and bottom waters. These movements of different water masses throughout the ocean are of great interest to oceanographers, as well as marine biologists, who are interested in how these movements affect the distribution of marine organisms.

We can use the T-S-D diagram to predict what may occur when two water masses of different densities meet. For example, a water mass with temperature 8°C and salinity 33.5 ppt intercepts another water mass with salinity 35 ppt and temperature 16°C).

1. What is the density of the first water mass? ___________ g/cm^3
2. What is the density of the second water mass? ___________ g/cm^3

When these two water masses meet, do you think they mix readily, or does something else happen? Explain:

When seawater freezes into ice (during winter months in the Arctic and Antarctic Oceans) salt is "squeezed out" of the ice crystals that form, and the salt is released into the surrounding unfrozen seawater, increasing its salinity. What do you think happens to this cold, high-salinity water after it is formed?

Concentration of Dissolved Gases:

A final property of water is its ability to contain various amounts of dissolved gases. Seawater contains dissolved CO_2 as well as dissolved oxygen (DO). All **heterotrophic** marine life depends on an adequate amount of DO in seawater, which for most organisms is at least 4 parts per million (ppm), equivalent to 4 mg of dissolved oxygen per liter of seawater. Oxygen is added to seawater by diffusion across the water-air boundary at the ocean surface, and also by photosynthetic organisms, such as unicellular phytoplankton and multicellular seaweeds and marine plants. DO is removed from seawater by the biological process of **cellular respiration** by marine organisms. Therefore, the amount of DO in seawater is highest in surface waters, but decreases with increasing depth below the **photic zone** to a critical depth known as the **oxygen minimum zone**.

Think of an unopened can of your favorite soda; it contains a certain amount of dissolved carbon dioxide (CO_2) gas under pressure which gives the soda its characteristic carbonation or bubbliness. When you open the can, the decrease in pressure causes the dissolved gas to come out of solution, forming bubbles of CO_2. In this case, the solubility of CO_2 (i.e., its ability to remain dissolved in liquid) decreases when the pressure decreases.

Temperature also affects the solubility of dissolved gases in a liquid. If you leave an opened can of soda out on a tabletop in a warm room, what happens to the level of carbonation in the soda after several hours?

Does the solubility of dissolved CO_2 increase or decrease as the soda warms up?

If a chilled can of soda is opened and placed into a refrigerator for a few hours, will the level of carbonation in the cold soda be greater or less than the first can, the one at room temperature, after the same amount of time?

Therefore, as you ______________________________ the temperature of a liquid, you ___________________________ the amount of gas that stays dissolved in the liquid.

Is this a direct (positive) or inverse (negative) relationship? ______________________________

This relationship is true for seawater in the oceans: as solar energy heats the water, oxygen and carbon dioxide molecules in seawater absorb thermal energy and become more mobile, which makes the molecules easier to escape as a gas from the water. Consequently, the amount of dissolved gases remaining in the seawater is reduced.

Salinity has a similar effect on the concentration of DO in seawater. The relationships between these 3 factors are illustrated by a **nomograph** (also called a Temperature-Salinity-O_2 solubility diagram, Figure 4). The nomograph shows the theoretical maximum amount of O_2 that can be dissolved in water of a given temperature and salinity, although actual samples of seawater may contain a higher concentration of O_2 (due to the photosynthetic activity of phytoplankton) or a lower concentration than expected (due to consumption of oxygen by marine organisms such as zooplankton).

To use the nomograph, lay a ruler or the edge of a sheet of paper across the axes to form a line connecting the values of salinity and temperature determined for your sample. The point where the edge intersects the middle line is the theoretical maximum amount of O_2 (in mg/liter) able to remain dissolved in seawater at that given temperature and salinity.

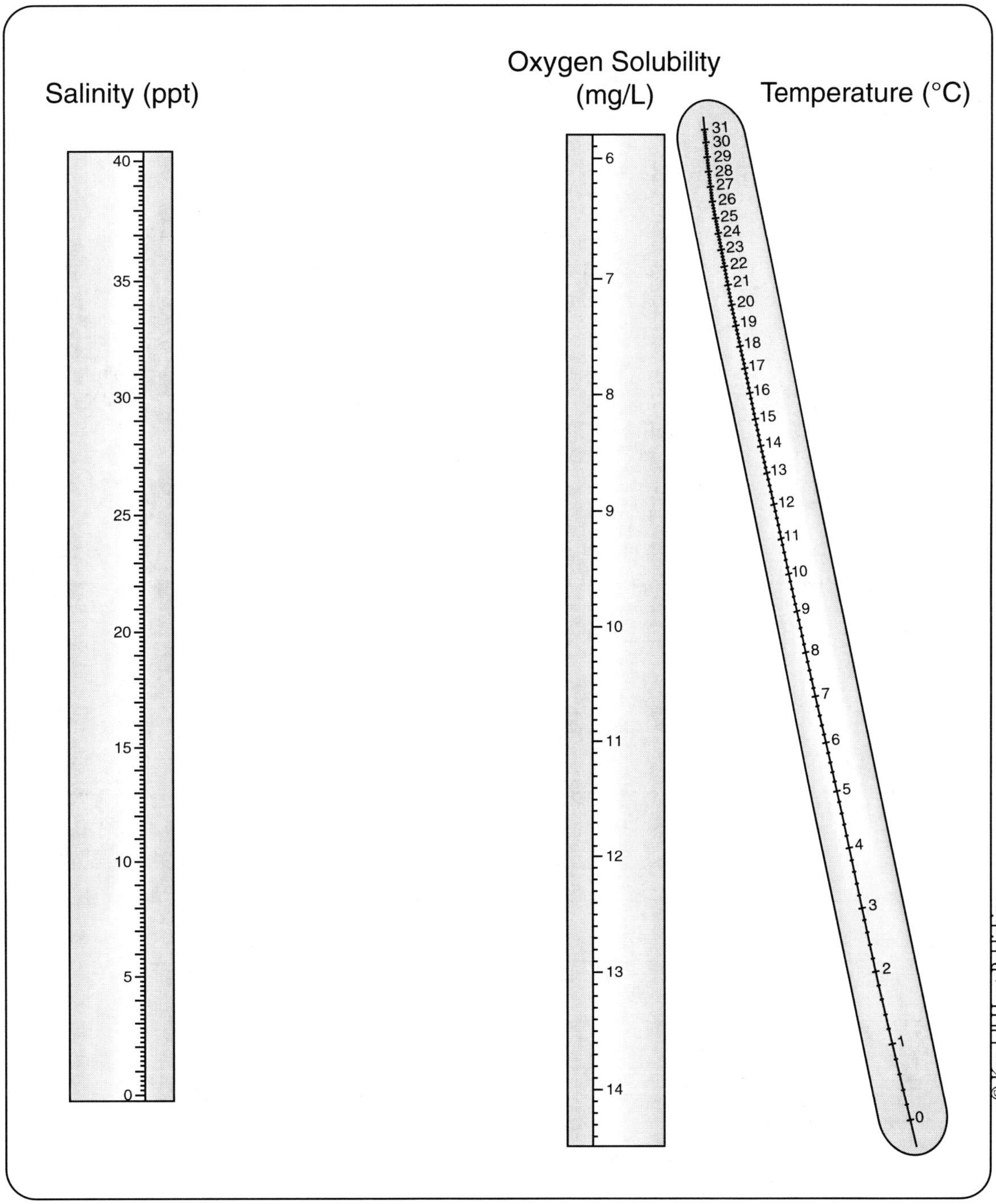

Figure 4. Nomograph showing the influence of temperature and salinity on dissolved oxygen content in water. If you know the values of two of the variables, a straight line connecting them will intersect the corresponding value of the third variable.

Other Methods of Determining Oxygen Content:

A standard procedure for the analysis of oxygen content is the **Winkler titration**. It involves a number of steps to chemically stabilize the free oxygen in the water sample; then acid is added to the stabilized sample and a chemical is added gradually until the sample changes color (when this occurs, it is called the endpoint). The amount of chemical used to reach the endpoint is proportional to the oxygen content at the time of sample collection. While this method is very accurate, it takes more time and hazardous chemicals are involved; because of this, it is more suitable to perform this procedure in the laboratory, not the field.

A more convenient and quicker way to determine oxygen content directly in the field is to use a **DO meter**. This tool measures an electric current produced when dissolved oxygen diffuses across a permeable oxygen-sensitive membrane housed within a sensing probe. The diffusion current detected by the sensor is directly proportional to the amount of oxygen in the water sample and is converted to units of concentration (mg/liter). Some DO meters also record temperature. If a DO meter is available for your use, your teacher will provide instructions for its operation.
Colorimetric methods. Simple-to-use chemical kits are available for testing oxygen levels in saltwater aquariums. These typically involve adding chemicals to the sample that change its color in proportion to the amount of oxygen in the sample being tested. The color of the sample is compared to a standard chart with colors that are calibrated to known levels of oxygen, and a match is determined visually.

1. Use any of the available methods to measure the temperature, salinity, and DO content of the 3 unknown water samples provided by your instructor, and enter this data in Table 5.
2. Use the nomograph to determine the maximum amount of oxygen that can remain dissolved in each sample at its measured salinity and temperature, and enter this data, as above.
3. Calculate how saturated your sample is with respect to dissolved oxygen. Express this as a percentage by using the equation below:

% Oxygen Saturation of Sample = (DO content as measured in Step #1 ÷ DO reading from nomograph) x 100.

Table 5. Properties of unknown water samples.

Your group data:

Sample #	Temp (°C)	Salinity (ppt)	Density (g/ml)	DO Measured (mg/L)	DO Theoretical (mg/L)	% Saturation	Identity of Sample:
1							
2							
3							

Class average:

Sample #	Temp (°C)	Salinity (ppt)	Density (g/ml)	DO Measured (mg/L)	DO Theoretical (mg/L)	% Saturation	Identity of Sample:
1							
2							
3							

Compare your calculations to the class average. How are yours different, and what are some reasons that could explain this?

Your instructor will reveal the identity of the water samples after you complete the exercise. After you learn the identity of your water sample, explain why you observed the results you did.

Name: ____________________
Date: ____________________
Section: ____________________

PROPERTIES OF SEAWATER REVIEW:

For questions 1–4, refer to the T-S-D diagram (Fig. 3).

1. What is the density of seawater with a temperature of 10°C and a salinity of 35 ppt?

2. What is the salinity of water that has a density of 1.0240 and a temperature of 20°C?

3. If temperature is held constant, how does seawater density change as salinity increases?

4. If salinity does not change, how does seawater density change as temperature changes?

5. In Exercise 4, you determined the salinity of a solution which was able to support an egg. What was the density of that solution when the egg reached neutral buoyancy?
   ________________________ g/cm^3

6. What was the density of the egg at that point? ________________________ g/cm^3

7. Complete graphs a–c by drawing a line showing the correct relationship between the variables. State whether each graph shows a *positive* or *inverse* relationship.

a.

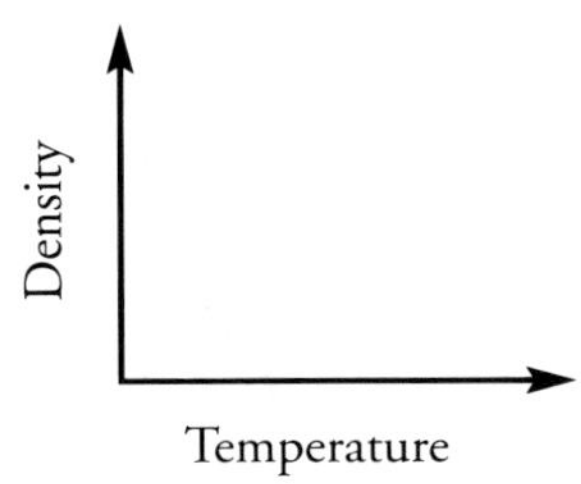

b.

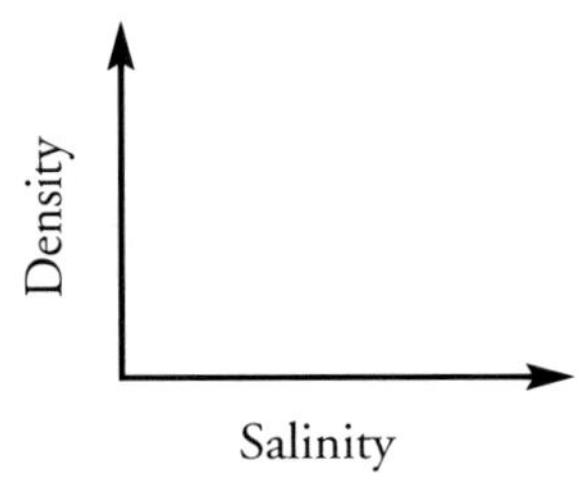

c.

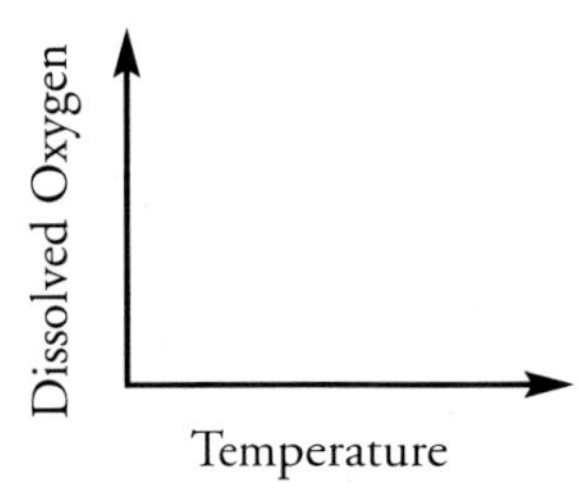

8. Imagine a neutrally buoyant, planktonic organism, drifting in the Mediterranean Sea (salinity 38 ppt and temperature 17°). How would its buoyancy change if currents carried it into the Atlantic Ocean (salinity 35.8 ppt and 19°C)?

9. In which environment is it easier to swim and float; the ocean or a swimming pool? Why?

10. Discuss the ways that different factors (both biotic and abiotic) affect the level of dissolved oxygen in seawater. Provide examples of at least 4 different factors in your answer.

WEB EXPLORATIONS FOR FURTHER STUDY:

Lamont-Doherty Earth Observatory Climate Data Library
http://ingrid.ldeo.columbia.edu/docfind/databrief/intro.html
The International Research Institute for Climate and Society (IRI) uses a science-based approach of strategic and applied research, education, and capacity building to enhance understanding of global climate issues and improve human welfare, especially in developing countries.

This website lets you access LDEO Climate Data and interpret images of global ocean patterns in temperature, salinity, and oxygen content.

- **Activity:** Scroll down to the section marked *Ocean Climatology,* and click on the world map next to it. This will bring up a data set in a new window. Scroll down to the area below the color bar and you will see a pulldown menu bar that reads *temp* (for temperature). Change temperature to *sal* (salinity), and click on the redraw button to the left of the map. A new page will load, and you will see a new map of the globe, showing salinity values. You can zoom in on any location by dragging a box around the area you want to inspect; a new window will open automatically.

  Find 3 seas with surface salinity higher than the average surface salinity of the world ocean (34 ppt) and list them in the table below. Next, find another 3 seas with lower-than-average surface salinity. Explain what may be causing the differences seen. Repeat this process to determine temperature and oxygen values of the surface water at the same locations you have selected. How does oxygen content compare between these regions? Is there a relationship between surface salinity and DO content, or temperature and DO content? Discuss your findings on the Web Explorations Report Sheet.

Location:	Salinity (PSU)	Temperature (°C)	Oxygen (mg/l)

- **Activity:** Research the natural history of all the following marine organisms, and complete the table. Then select four of the planktonic animals listed below and report on their unique adaptations that govern their buoyancy.

Common Name:	Scientific Name:	Phylum:	Lifestyle: (planktonic, benthic, or nektonic)	Type of Buoyancy:		
				Positive	Negative	Neutral
Portuguese Man-o-war	*Physalia physalis*					
Great white shark	*Charcharodon carcharias*					
Salp chain	*Pyrosoma sp.*					
Moon Jellyfish	*Aurelia aurita*					
By-the-wind-sailor	*Velella velella*					
Diatoms	*Chaetoceros sp.*					
Blue sea slug	*Glaucus atlanticus*					
Pacific halibut	*Hippoglossus stenolepis*					
Sunfish	*Mola mola*					

- **Activity:** Prepare a report on **thermohaline circulation** and **downwelling** throughout the ocean basins. Describe these processes, and the role photosynthetic organisms play. Explain why these circulation patterns are critically important for life in the deep sea.

Name: ____________________
Date: ____________________
Section: ____________________

WEB EXPLORATION REPORT

Topic: ____

LABORATORY 3

Classifying and Identifying Marine Invertebrates

Scientists estimate there may be as many as 100 million species of living things on Earth, though slightly over 1.75 million have been formally described. The deep sea undoubtedly harbors many as-yet undiscovered species, so as scientists continue to explore the sea's depths using submersibles and remotely operated vehicles, the next generation of marine biologists will certainly encounter a huge variety of bizarre new life forms to study. Biologists will deal with all of this diversity by organizing those new species into logical groupings, based on similar anatomical features. This is the basis of **taxonomy**, the science of classifying life. In organizing species into groupings, the goal is to show how the organisms are related to each other through evolution, to show what is thought to be their evolutionary history (their **phylogeny**). Taxonomy is a constantly-changing field, as research in molecular biology reveals DNA similarities that indicate new and closer relationships between species.

One of the most important skills for a marine biologist to have is a broad knowledge of the diversity of oceanic life and an understanding of its classification so as to recognize which particular **taxon** (plural **taxa**) an organism belongs to. This skill is fundamental for identifying a wide range of organisms.

The modern system of classification is based on a hierarchy of taxa, with the largest group being the Kingdom. The Kingdom Animalia is subdivided into **Phyla** (singular, **Phylum**); presently there are between 35–40, depending on the classification system being used. Phyla are divided into Classes, and so on as shown:

All these together are 7 **taxa**

Kingdom
Phylum
Class
Order
Family
Genus
Species

This is one **taxon** (→ Family)

This hierarchy places organisms showing a large number characteristics into **genera** (singular, **genus**), with the **species** being the most specific and inclusive grouping; all organisms in this taxon share nearly identical characteristics.

For example, let's classify our species, *Homo sapiens:*

Taxon Name:	Features of this taxon:
Kingdom Animalia	these organisms ingest food, are multicellular, and have a larval form
Phylum Chordata	dorsal nerve cord and a stiff rod (the notochord) extending the length of the body
Class Mammalia	hair, mammary glands, and endothermic metabolism
Order Primates	opposable thumb, nails instead of claws
Family Homonidae	bipedal, walking erect
Genus *Homo*	
Species *sapiens*	(other species, *H. habilis, & H. erectus,* are extinct)

A species is identified by its scientific name, which is the genus and species given together, like giving a surname first then a given name. The genus name is capitalized, but the species never is, and scientific names are always *italicized* or underlined, so the proper scientific name of a human being is *Homo sapiens* or Homo sapiens, but never Homo sapiens or *Homo Sapiens.*

Let's compare our classification to that of the sea otter.

Taxon Name:	Features of this taxon:
Kingdom Animalia	Here we see that humans share characteristics of these 3 taxa with the sea otter, but at the level of Order, we diverge.
Phylum Chordata	
Class Mammalia	
Order Carnivora	well-developed canine teeth
Family Mustelidae	anal scent glands (though absent in the sea otter)
Genus *Enhydra*	webbed hind feet
Species *lutris*	

A weasel has more traits in common with a sea otter, since both are in the family Mustelidae; thus, weasels are more closely related to otters than humans are. In evolutionary terms, we would say weasels and sea otters shared a common ancestor more recently than primates and otters did. Another way to say this is that the lineage that included the common ancestors of both primates and otters diverged a much longer time ago than the lineage that contained the common ancestors of weasels and otters.

Since there are so many different kinds of organisms, biologists use identification guides called **keys** to help them identify organisms quickly and easily. These keys are usually based on recognizing external structures on organisms, so careful observation is critical. Most keys are written to help identify a certain taxon or organisms within a certain geographic region or habitat.

The simplest kind of key is **dichotomous** (Greek: *dich* = two, *temnein* = to cut), providing a pair of questions (or couplets) to answer about the organisms under scrutiny. If the question can be answered "yes," then you are directed to answer another couplet. If the question is false or can be answered "no," then you proceed with the next questions as they are listed in the key. In this way, you narrow down the choices until you reach the name of the taxon or the organism that you are trying to identify.

For example, here is a diagram of a how a dichotomous key subdivides 6 kinds of produce.

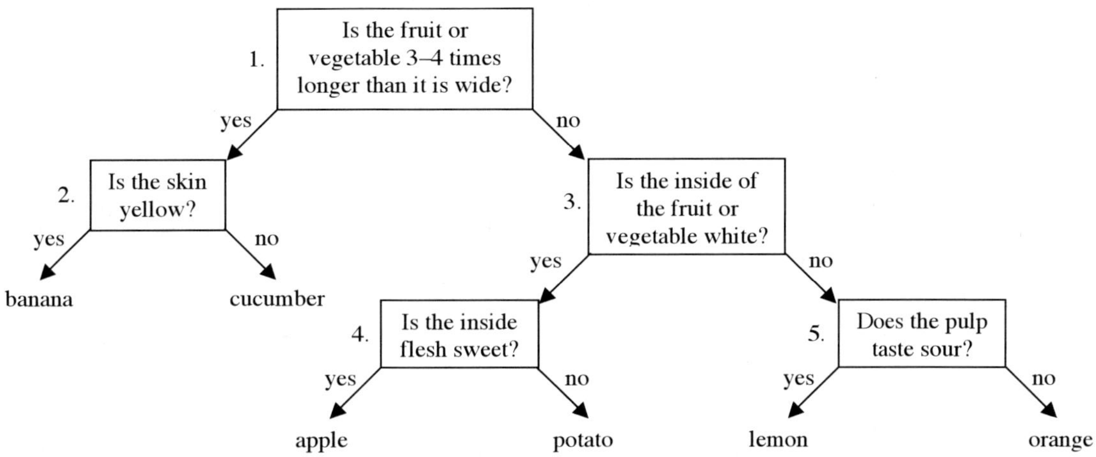

Figure 1. Flow chart showing how "yes" or "no" decisions lead to identification of specimens in a dichotomous key.

Here is how the information displayed above is organized into a key:

1. a. Fruit or vegetable is 3–4 times longer than it is wide 2
   b. Fruit or vegetable is not longer than it is wide 3

2. a. Skin is yellow .... banana
   b. Skin is not yellow .... cucumber

3. a. Inside of the fruit or vegetable is white 4
   b. Inside of fruit or vegetable is not white 5

4. a. Inside flesh is sweet .... apple
   b. Inside flesh is not sweet .... potato

5. a. Pulp tastes sour .... lemon
   b. Pulp is not sour .... orange

Notice that there is always one couplet fewer than the number of items being identified.

I. EXERCISE IN CREATING A KEY

Your instructor will provide you with several kinds of seashells. Study their morphology closely, using Figure 2 as a guide, and determine an anatomical feature or characteristic that could be used to divide the specimens into two groups.

1. Record the trait that made the division, and separate the shells into two groups: a YES group (the shells that have the trait) and a NO group (shells lacking the trait).
2. On a sheet of scratch paper, make a flow chart diagram showing this division (as shown in box 1 of Figure 1). Divide the YES group by using another single characteristic and represent this on your diagram (as shown in box 2 of Figure 1). Similarly, if needed, divide the NO group into two groups, and show this in your flow chart.
3. Continue this dividing process until each shell is by itself.
4. When you have separated all the shells according to this technique and are satisfied that your flow chart works accurately, convert the chart into a dichotomous key with couplets, as shown in the preceding example. Use **quantitative** measurements rather than **qualitative** terms of size: say "shell is 5-6 cm in length" rather than "shell is medium-sized." Use the terms in Figure 2 and the definitions below.
5. When your key is complete, give it to another student and see if he or she can accurately identify your shells.

Anterior (or siphonal) canal – opening at the base of the shell aperture.
Aperture – cavity in the shell where the animal's body fits.
Apex – the tip of a spiral shell.
Axial ribs – raised lines on a shell that extend from the apex to the anterior canal, crossing the whorls. On a bivalve, these lines run from the umbo to the edge of the shell.
Bivalve – a shell composed of two valves connected by a hinge.
Body whorl – the largest whorl, which houses the body of the animal.
Concentric ornaments – grooves that extend from the umbo in parallel curved lines.
Spire – all the whorls above the body whorl, to the apex.
Suture – a groove that separates two whorls.
Umbilicus – a cavity at the base of a spiral shell.
Umbo – the oldest part of a bivalve shell, above the hinge.
Univalve – a one-piece shell.
Valve – one shell of a bivalve.
Whorl – one individual turn or coil of a spiral shell.

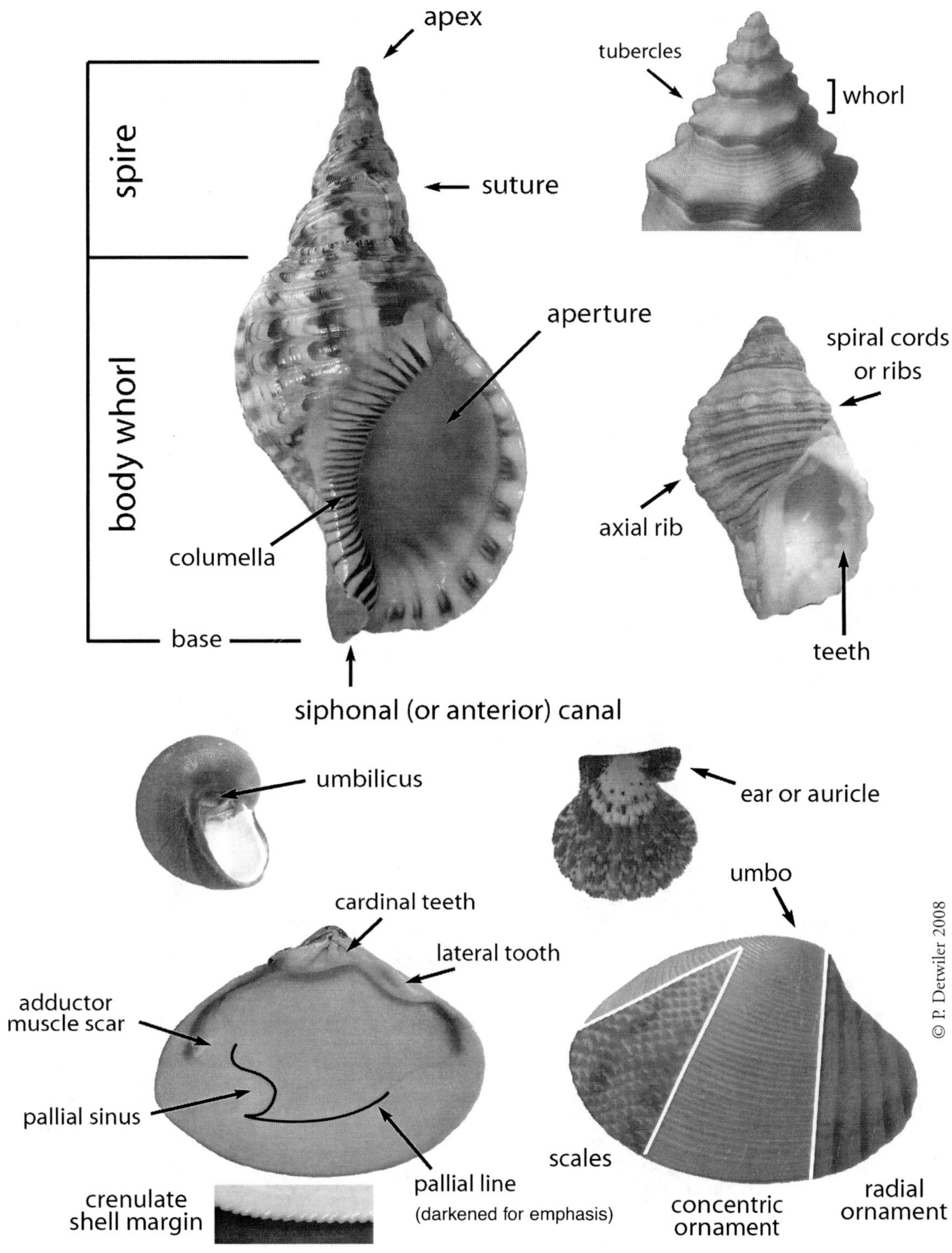

Figure 2. Illustrated terms used to describe the external features of mollusk shells.

Did your key lead another user to the same identifications? If there were problems, review the characteristics you chose to separate the groups. Did the other user misunderstand your descriptions, or were your descriptions not specific enough? Analyze the results and if necessary, modify your key to eliminate confusion.

Why are scientific names important? Why not just use common names to identify organisms? (Remember, science is an international discipline.)

II. KEY TO MAJOR PHYLA OF MARINE INVERTEBRATES

In this exercise, you will use a simple key to identify common marine organisms and classify them into phyla. Because of the great diversity in body forms of marine invertebrates within each phylum, you may not be able to identify every specimen set before you. This key is meant to be a general "first pass" at identification. Your instructor may also provide you with more detailed keys that are specific for organisms in your geographical region.

Procedure:

1. Examine the numbered specimens provided to you. Try to identify which body structures are used for the following life processes:
   a. Reproduction
   b. Eating and prey capture
   c. Respiration and gas exchange
   d. Locomotion
2. Use the following key to identify the specimens, and fill in the following data table with your identifications.
3. Make a sketch of each numbered specimen in the following boxes, and label any body parts or structures that are unique to animals in that phylum.

Key to the common macroscopic marine invertebrates found in coastal regions.

1. a. Numerous pores and holes perforate the body, texture feels spongy or fibrous .... Phylum Porifera
   b. Not as above 2

2. a. Colonial organism, individual members of the colony (**zooids**) are embedded in rubbery material or in tiny square boxes 3
   b. Not colonial, or if colonial, individuals surrounded by a hard calcareous skeleton 4

3. a. Zooids in a star pattern or embedded in rubbery material ...... Phylum Chordata, tunicates
   b. Zooids in regularly shaped boxes .... Phylum Ectoprocta (bryozoans)

4. a. Body symmetry is **radial**, may have a circle of unbranched tentacles at one end 5
   b. Body symmetry is **bilateral** with no **pentamerous** structures 6

5. a. Jellylike body, unbranched tentacles surround a central slit-like mouth .... Phylum Cnidaria
   b. Body is **pentamerous**, or tentacles in multiples of 5s .... Phylum Echinodermata

5 long arms, radiating from a central disk ....brittle star
Arms tapering to the center...sea star
Spiny hemispherical body, flattened on one surface ....sea urchin
Body is flatter than it is wide..sand dollar
Body is sausage-shaped, with tentacles at one end ....sea cucumber

6. a. Body is very flattened and/or wormlike 7
b. Body not wormlike 11

7. a. Well-defined external segments 8
b. No visible segments...9

8. a. Bristles or flap-like structures are present on body segments ....Phylum Annelida
b. Bristles are absent from the body ....Phylum Platyhelminthes (tapeworms)

9. a. Body is very thin (< 1 mm) and flattened **dorso-ventrally**
(top to bottom) ....Phylum Platyhelminthes (flatworms)
b. Body not as above 10

10. a. Body is threadlike, round in cross section, both ends taper to a point ....Phylum Nematoda
b. Not as above 11

11. a. Body is covered by a hard **exoskeleton** with jointed **appendages** and
compound eyes...Phylum Arthropoda
Cephalothorax is wider than long, has claws ....crabs
Cephalothorax is longer than wide ....shrimp/lobster
Cephalothorax is flattened so body is taller than wide, legs are different shapes
and sizes ....**amphipods**
Cephalothorax is flattened top to bottom, no claws, legs are same size and shape......**isopods**
Eyes are fused to exoskeleton, has no antennae
Large cephalothorax, a single spine for a tail ....horseshoe crab
Small body, 8 long legs, all the same size...sea spider
b. Body not covered by exoskeleton, has single eyes 12

12. a. Animal either crawls with a large muscular **foot**, has 2 shells (**valves**),
or has tentacles and large eyes ....Phylum Mollusca
Sluglike body or body covered with one shell ....Class Gastropoda
Dorsal surface has 8 shell plates ....Class Polyplacophora (chitons)
Clamlike, two hinged shells enclose the body..Class Bivalvia
Tuft of filaments are visible at hinge of shell ....mussel
Upper and lower valves are symmetrical ....clam
Valves have ear or auricles at hinge ....scallop
2 large eyes and 10 appendages bearing suckers ....Class Cephalopoda
b. Not as above ....Phylum Chordata

Vocabulary for Invertebrate Key:

Appendage — Any body part arising from the main trunk of an animal, such as antennae, legs, or claws.

Bilateral symmetry — A pattern of body organization where the right and left sides of the body are externally identical.

Cephalothorax — The fused head and thorax region of many types of crustaceans; most conspicuous on lobsters.

Dorsal — Related to the upper body surface of an animal; the opposite of ventral.

Exoskeleton — The hard body covering of arthropods.

Foot — A large muscle in mollusks that has been modified in several ways: for burrowing in bivalves, for crawling in chitons and gastropods, and for grasping prey in cephalopods.

Pentamerous — Related to body symmetry based on 5 parts.

Radial symmetry — A type of body organization where all the body parts are arranged symmetrically around a hub to each other.

Valve — One of the calcareous halves of a clam, mussel, or other bivalve shell.

Zooid — A tiny sessile invertebrate that reproduces clonally to form a colony; each clonal individual in such a colony.

Table 1. Record of Identified Invertebrate Specimens.

Common name:	Phylum:	Class (if known):
1. Sea Urchin	Echinodermata	
2. Squid	Cephalopoda	
3. Brittle star	Echinodermata	
4. Sponge	Porifera	
5. Mantis Shrimp	Arthropoda	
6. Sea Anemone	Cnidaria	
7. Marineworm	annelid	
8 Shrimp	Arthropoda	
9. Red rock crab	Arthropoda	
10. Shrimp/lobster	Arthropoda	
11. Seaspider	Arthropoda	
12. Chiton	Mollusca	

1.

2.

3.

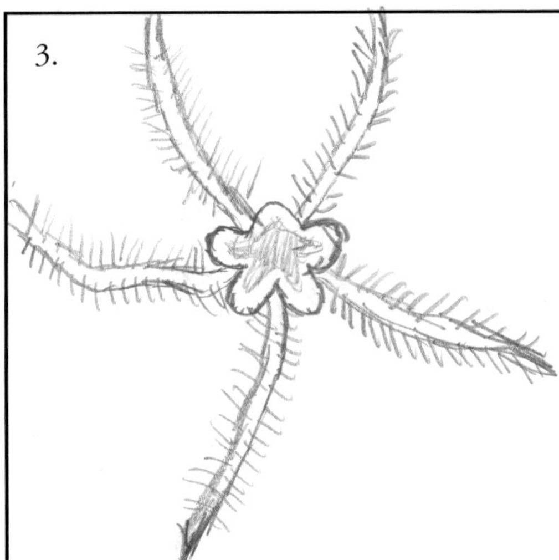

4.

5.

6.

7.

8.

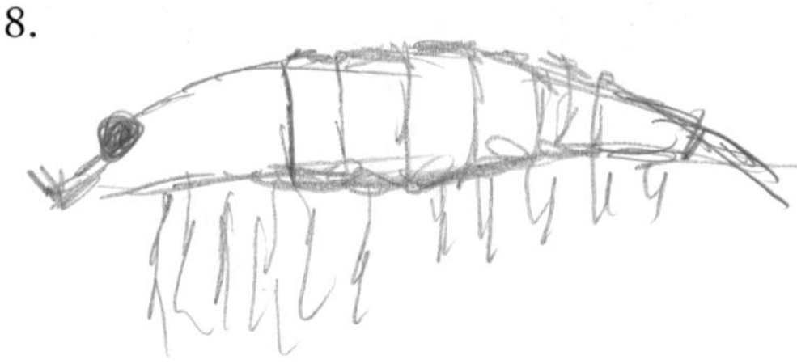

9.

10.

11.

12.

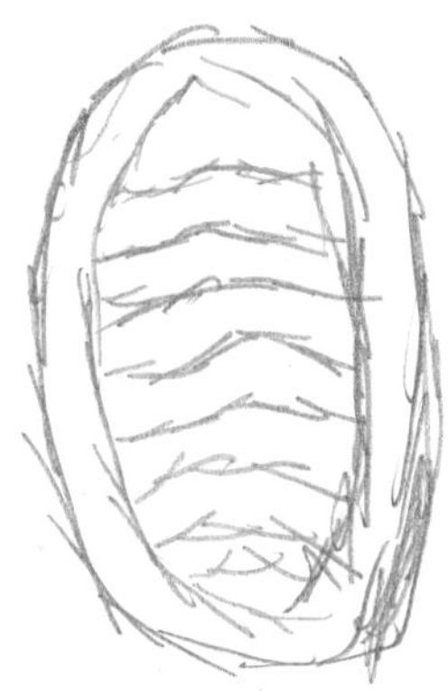

LABORATORY 4

An Exercise in Observing and Describing Marine Organisms

INTRODUCTION:

People who work in technical fields need to learn specialized vocabularies to help them communicate ideas and describe observations accurately and concisely to others, such as when marine biologists write research reports for governmental agencies and policymakers. Information must be transmitted in clear, direct statements with little chance of misinterpretation.

One way that scientists differ from non-scientists is the degree to which scientists use numbers to describe their observations. For example, while a nonscientist may say a room is full of people, a scientist making a scientific observation of the same room might say, "The room contains 6 men and 8 women." Similarly, instead of saying, "A *big* wave crashed on shore," a marine biologist might say, "A *3-meter* wave crashed on shore."

Scientists attempt to describe events and phenomena with a greater degree of accuracy by using **quantification** (the act of expressing the quantity of something using standard scales of measurement), rather than by using **qualification** (the act of expressing the characteristics or traits of something) alone.

It is important for biologists to develop focused observational skills and a keen attention to detail. Therefore, this exercise requires you to carefully examine the anatomical features of a marine organism (in particular, the surface features of a mollusk shell) and to communicate your observations to others as effectively as possible.

OBSERVATIONS:

Your teacher will give you a mollusk shell. Examine your shell closely. Can you make any predictions about the lifestyle of this mollusk species on the basis of the appearance of its shell? Try to answer the folllowing questions:

- In what type of habitat (bay, estuary, beach, rocky area) did the animal live?
- How much wave energy is typical of its environment (did the animal live in rough or calm water)?
- On what feature(s) of the shell's appearance do you base your predictions?
- Is your species **sessile**, remaining in one place, or is it mobile, crawling around on the sea bottom?
- How does the animal feed? Is it a predator, herbivore, or **filter feeder**?
- Can you tell this species' method of feeding from the shape or features of its shell?

Observations: When alive, was your specimen **epifaunal** (living on a firm substrate or on the surface of other animals) or **infaunal** (living in or within sediments or hard substrates)? How can you tell?________

Did your specimen live in a high-energy environment (meaning it was exposed to heavy surf and waves)? How can you tell?_____

Is your specimen a mobile or sessile species? How can you tell?______________________

Sit back-to-back with your partner. You have 5 minutes to describe the details of your shell to your partner, who will draw the shell in Box 1 in his or her lab worksheet. Do not show or reveal the identity of the shell to your partner! If you know what kind of shell you have, do not say its name. While your partner draws from your description, he must not show you his drawing until time is called.

> *Example:* Say you are given the shell of the common garden snail (*Helix aspersa).*
> Do not say, "This is a brown garden snail, the kind you see outdoors." Instead, describe it as a *1-2 cm diameter gastropod shell, coiling to the right, with a brown shell and white stripes, etc.*

Box 1.

Details of shell: Record notes on texture, color, markings, size, and other features that you notice on the shell surface.

After 5 minutes, change roles. Your partner will now describe to you his/her shell and you will draw it in Box 1 of your worksheet.

After both you and your partner have drawn each other's shells, your teacher will call time and distribute a handout to each of you to help you with the next drawing you will make. Your partner will use the handout to help him/her redescribe the same shell to you, but this time, using more specific terminology. Refer to your handout as you follow your teacher's instructions, and make your second drawing of your partner's shell as he/she describes it to you in Box 2.

Box 2.

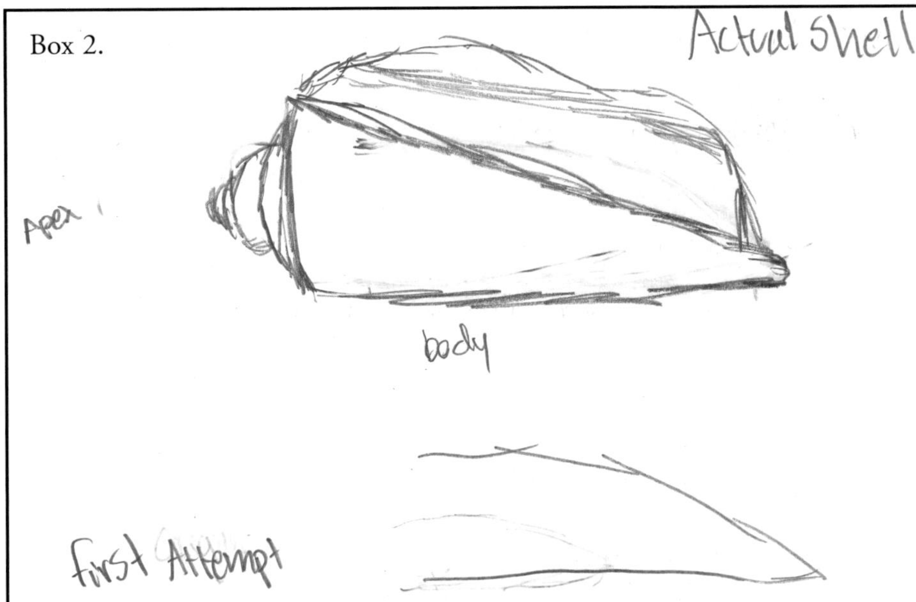

After time is called, change roles as before. After 5 minutes, compare the 2 drawings you both made to the actual shells you and your partner have.

In your first drawing, you probably were not familiar with the anatomical terms for mollusk shells, so you had to describe the shell to your partner using general, non-technical words that may have been more subjective and open to interpretation. Therefore, you and your partner may not have interpreted each other's descriptions accurately.

For the second drawing, when both of you used a common, specialized vocabulary with quantitative terms for describing the shell (and used the ruler as a standard to express size accurately), how did your drawing improve? Identify 3 ways:

1. The measurements were correct

2. The general shape was similar

3. Describing the small apex and siphonay

CONCLUSION:

What do you think were the purposes of this laboratory exercise? Discuss with your partner and come up with a list of 3 objectives that you think you have learned as a result of this lab. Write them below, and share them with the class.

1. To stress observation on a basic thing
2. Using terms help describe the shey
3. Communicating with a lab partner

Review: Scientists seek to emphasize accuracy of their observations by using words that describe measurable characteristics in a(n) Objective (*subjective/objective?*) manner. Additionally, they use terms that are more scientific and quantitive

Which statements are objective/subjective and qualitative/quantitative? Place a check mark in the appropriate box(es).

	Qual.	Quant.	Subj.	Obj.
The majority of students at the college attend school part-time.	✓			✓
Eighty percent of the students at the college work 20 hours or more per week.		✓		✓
That shade of green is prettier than the other color.	✓		✓	
At sea level, pure water boils at 100° Celsius.		✓		✓
Fifteen percent of 60 is 9.		✓		✓
The shell of the Queen conch (*Strombus gigas*) is large, pointed, and heavy.	✓		✓	
Our school has a good football team this year.	✓		✓	
Our school's football team has a 15-0-0 record this year.		✓		✓
There are 10 excellent orators and 5 terrible ones on the debate team this year.	✓	✓	✓	

EXTENDING YOUR KNOWLEDGE:

Compare a different shell to the one you previously described in this exercise. Describe 5 features that differ, using proper terminology from the handout that your teacher provided.

- It has a smaller apex
- The siphonal is wider
- The aperture has more access
- It lacks ribs since it's smooth
- It has a smaller suture

LABORATORY 5

Marine Microbiology

INTRODUCTION:

Since the late 1990s, new technologies have allowed marine microbiologists to make numerous groundbreaking discoveries about the world of microscopic marine organisms (*Bacteria*, viruses, *Archaea*, and fungi). The roles that these tiny microbes play in the global ocean ecosystem are extremely important due to their sheer numbers and the biomass they contribute to the oceanic food chain. Only recently have their contributions to global-scale biochemical processes begun to be quantified. New methods of analysis reveal that a single drop of unpolluted seawater may normally contain as many as 10^6 (1,000,000) bacteria cells and perhaps 10 times that number of viruses. Fortunately, the vast majority of these are not **pathogenic** (disease-causing). On a global scale, the activity of these organisms in the turnover of nutrients and conversion of organic matter (whether they act as consumers or decomposers) is massive, and although these life forms are normally invisible to us, they are nonetheless indispensable to the health of the world ocean.

Over the course of several lab periods, you will study a variety of bacteria found in seawater and from different marine sediments. You will also attempt to culture luminescent (light producing) bacteria which are found as free-living forms throughout the world ocean in seawater and as **symbionts** within the bodies of many types of marine organisms. Because this light is produced through biological processes by living organisms, it is called **bioluminescence**.

PROCEDURES (DAY 1)

Wash your hands thoroughly and clean the work surface of your lab bench using paper towels and disinfectant. Why is this first step always important whenever working with bacteria?

Preparation of Seawater Plates:

1. Use a pipette to transfer 0.1 ml of seawater onto a plate of agar media.
2. Sterilize a glass spreading rod as shown by your instructor. Use it to evenly distribute the seawater sample over the entire surface of the agar.
3. Cover the plate with its lid and let it sit for 5 minutes to allow the agar to absorb the sample.
4. Invert the plate and label the agar side with your initials, the words "Seawater R1," and the date. Tape the lid onto it. The plate is stored with the agar side facing up.
5. Repeat Steps 1–4 to make a replicate plate, except label it "Seawater R2." The plates will be incubated at 35°C or room temperature until the next lab period.

Preparation of Sediment Plates:

You will study sediment samples collected from two coastal locations differing in the amount of daily wave exposure. One sample was collected from a sheltered area, such as a bay or estuary, and the other from a location on the open coast.

1. Fill a sterile test tube 1/3 full with sterile seawater and add the same amount of sediment from the bay/estuary location. Stopper the tube and shake it steadily for about 20 seconds to suspend the bacteria in the seawater. Place the test tube into a rack and let the sediment settle.
2. Use a pipette to transfer 0.1 ml of the liquid onto a plate of agar media.
3. Use a glass spreading rod to evenly distribute the sample over the entire surface of the agar. Sterilize the glass spreading rod as shown by your instructor.
4. Cover the plate with its lid and let it sit for 5 minutes to allow the agar to absorb the sample.
5. Invert the plate and label the agar side with your initials, the word "Sheltered," and the date. Tape the lid onto it. The plate is stored with the agar side facing up. Repeat Steps 1–4 with sediment from the other location, but label it "Exposed." The plates will be incubated at 35°C or room temperature until the next lab period.

Bacteria are decomposers. Which location might harbor more bacterial decomposers? Why?

On which plate would you predict to see more bacterial growth?

With your partner, form a hypothesis that you can address with this study.

PROCEDURES (DAY 2)

Growth of Colonies Found in Seawater and Sediment Samples:

At the next lab period, examine your plates. Look at the morphology of at least 7 of the colonies that formed, and note color, size, shape (elevation), shape (of colony margin), and surface appearance. Refer to the following illustrations in Figure 1 for descriptive terms to use. Use a grease pencil to draw a circle on the upper lid of the Petri dish. directly above each colony. Number these colonies 1 through 7 on the lid. Record your observations (for both the seawater and sediment samples) in your datasheet.

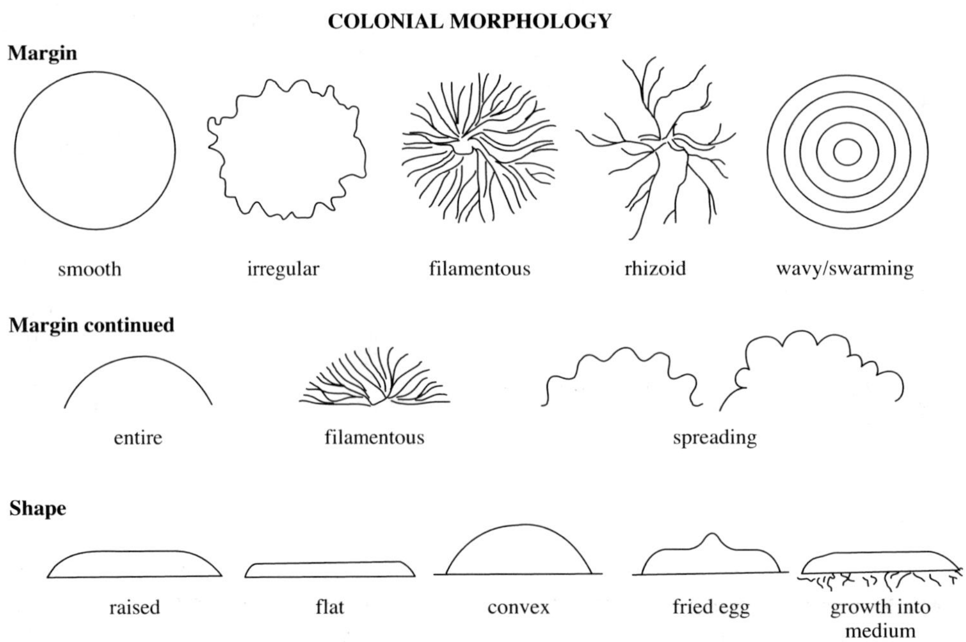

Figure 1. Illustrations describing morphology of bacterial colonies grown on agar.

Examination of Seawater Plates for Bioluminescence:

Take both seawater plates into a completely dark room and, after your eyes get used to the dark, look for luminescent colonies. If any are present, describe the color of the bioluminescence on the datasheet. Record the position of any luminescent colonies by circling them on the lid of the sealed Petri plate with a marking pen. After you have returned to a lighted room, mark those same colonies by circling them on the bottom surface of the Petri plate (the side containing the agar).

Total number of colonies on plates: ____________
Number of colonies that are luminescent: ____________
Percentage of colonies that are luminescent: ____________

The next step is to isolate any luminescent bacteria and grow a pure culture, as described below.

DVD:
Watch the video demonstration for preparing a streak plate several times, so that you are familiar with all the steps of the procedure before you begin.

Isolation of Bacteria Using the Streak Plate Technique (Overview):

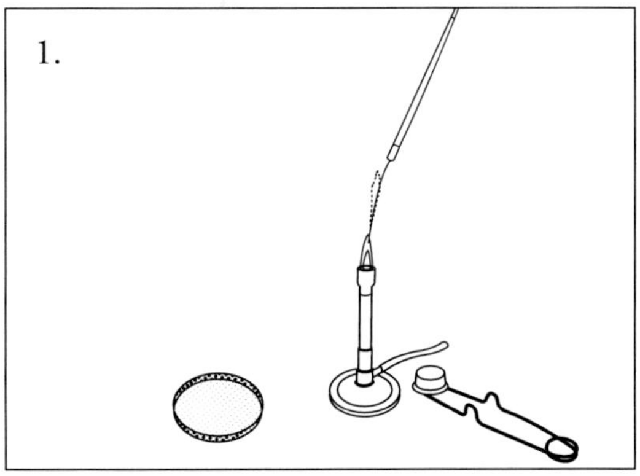

After flaming and cooling a wire loop, pick up a loopful of the culture or a portion of a colony from a fresh plate.

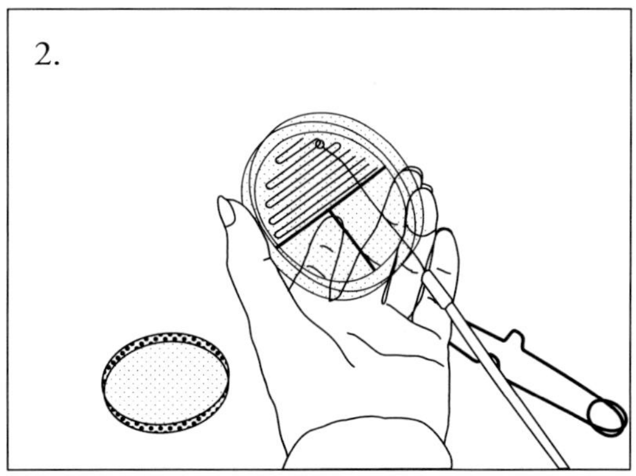

Starting on one side, streak the primary area by streaking about 1/3 of the plate. Flame and cool the loop.

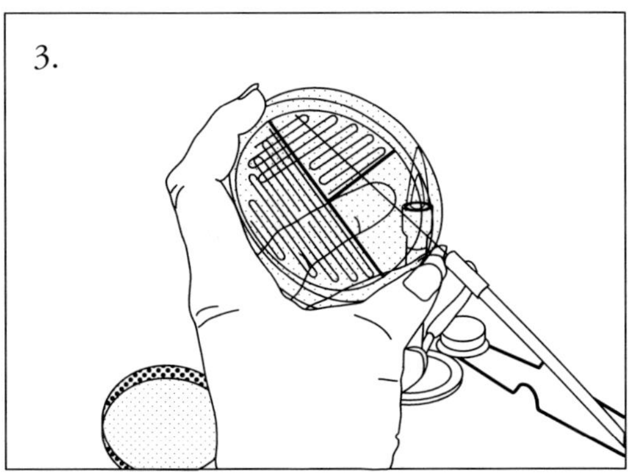

Turn the Petri plate about 1/4 turn and streak the secondary area by going back into the primary area only once or twice and then streaking four to five times in the secondary area. Flame and cool the loop.

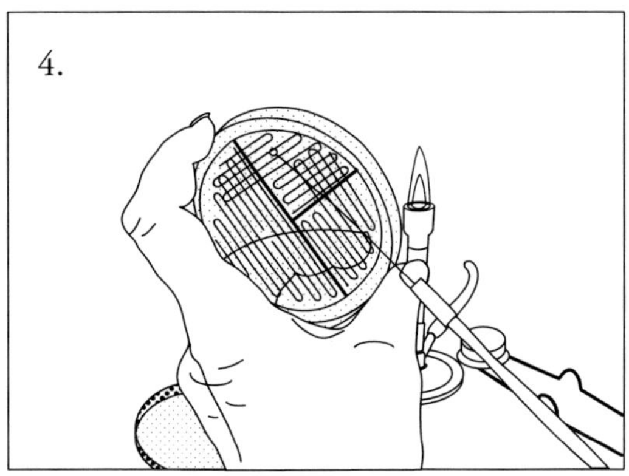

Turn the plate another quarter turn and continue streaking by passing into the secondary area once or twice and then streaking out to the end of the plate.

Step-by-step instructions for the streak plate technique are provided below.

1. Light a Bunsen burner or other flame source.
2. Flame a wire inoculation loop to sterilize it (hold the tip of the wire in the flame until the wire glows cherry red, then withdraw it and allow to cool).
3. Remove the lid from the Petri plate containing the seawater bacteria.
4. Use the sterilized wire loop to pick up a loopful of bacteria from a luminescent colony. Replace the lid.
5. Remove the lid from a new, sterile seawater/agar media plate, and streak 1/3 the plate. The wire should skim across the surface of the agar and not dig into the agar.
6. Cover the new plate with its lid. Flame and cool the loop.
7. Remove the lid, and streak 1/4 of the plate by going back into the first streaked area once or twice and streaking out into the new area.
8. Cover the plate with its lid. Flame and cool the loop.
9. Remove the lid, and streak the remaining 1/4 of the plate. Flame the loop and set it aside.
10. Replace the lid on the plate and label the agar side with your initials, the word "Biolum," and the date. Tape the lid onto it. The plate is stored with the agar side facing up. The plate will be incubated at 35°C or room temperature until the next laboratory period.

In order to view bacterial cells (which are transparent), they need to be stained with a dye. First the cells are transferred onto a slide (and fixed), and then the cells are stained. Follow the procedures below to prepare and stain a bacterial smear.

DVD:
Watch the video demonstrating how to prepare and stain a bacteria smear. Also review the video segment on Oil Immersion technique in the Microscopy chapter of the DVD.

Procedure to Make a Smear Preparation:

1. Label a microscope slide with the name of the plate and the number of the bacterial colony you are sampling.
2. Draw a circle on the underside of the slide with a grease pencil.
3. Using a sterile loop, place a small drop of water within the circle.
4. With the same loop, pick up a tiny amount (barely visible to the eye) of the bacterial colony and stir it into the water drop on the slide.
5. Allow the slide to air dry. Do not heat the slide in a flame to speed up the drying process; this will rupture the bacteria cells.
6. When the slide is completely dry, pick it up with a wooden clothespin and heat fix the slide by quickly passing its bottom surface through a flame 3-4 times, as shown in the DVD. This heating process kills the bacteria and adheres the cells onto the slide so they will not wash off during the following staining process.

SMEAR PREPARATION PROCEDURE

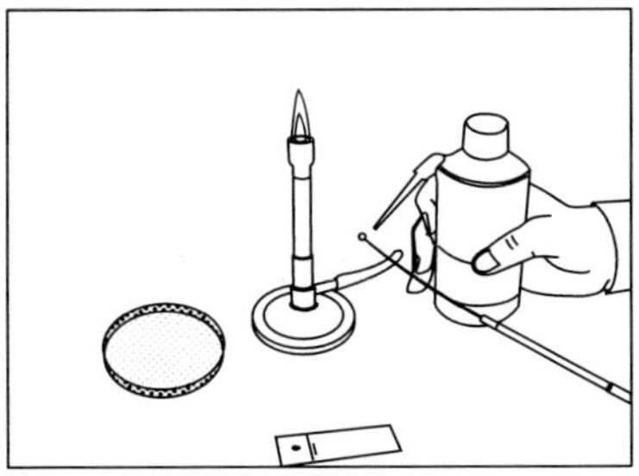

1. After flaming and cooling the loop, add a drop of water to the loop.

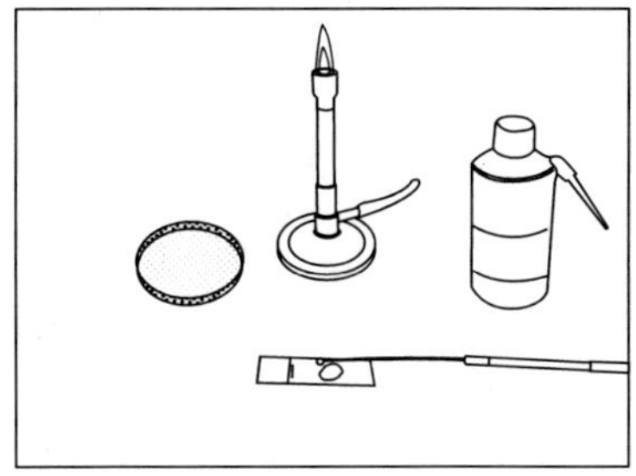

2. Draw a circle on the bottom of the slide, and place the water droplet on the microscope slide in the area of the circle.

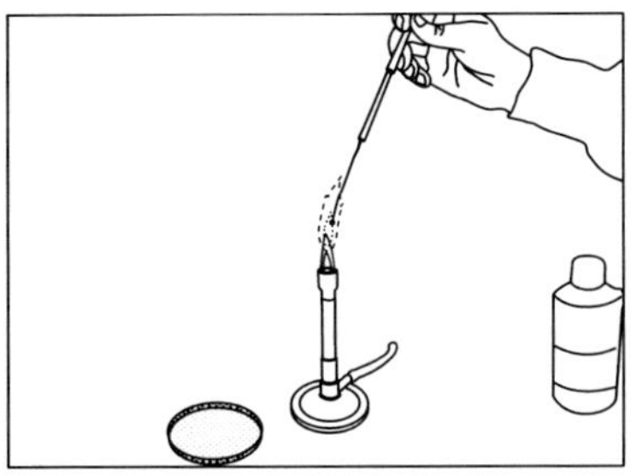

3. Flame and cool the loop.

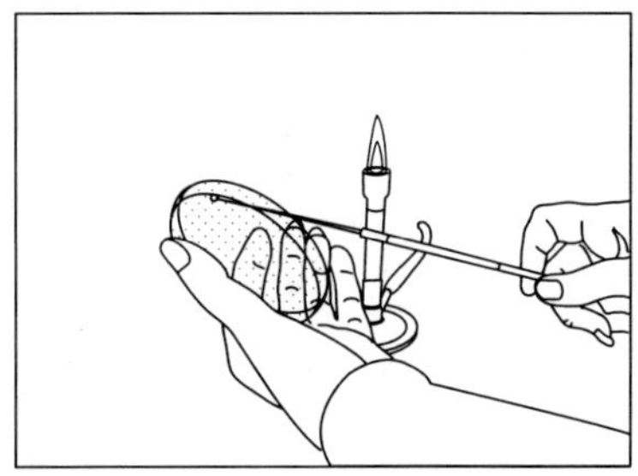

4. Pick up a small amount of culture from an isolated colony on a plate.

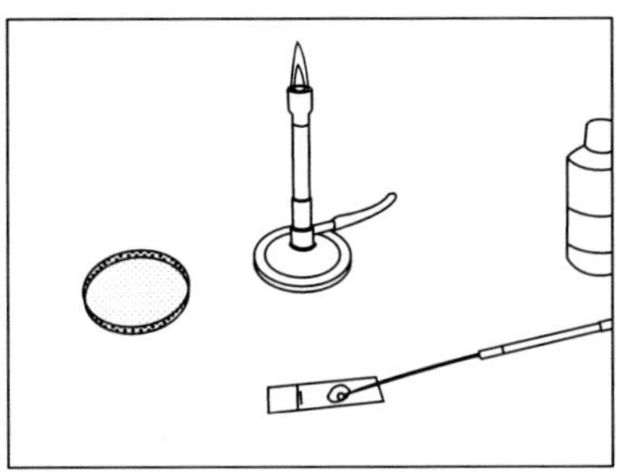

5. Spread the loop of bacteria in the water droplet in the area of the circle.

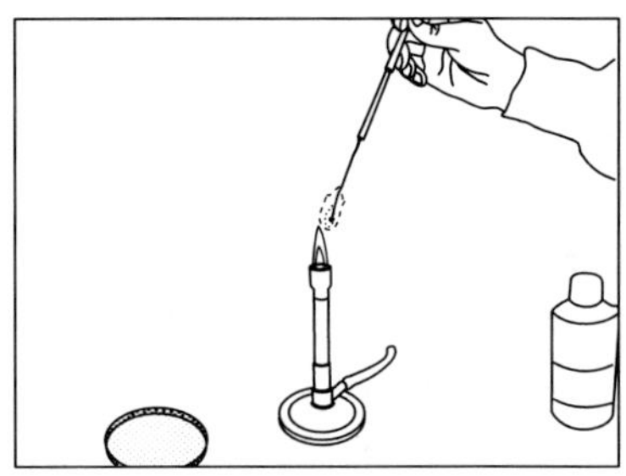

6. Flame and cool the loop.

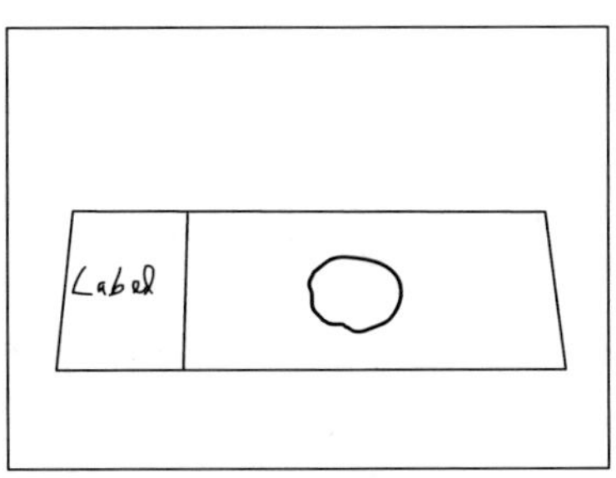

7. Air dry the slide.

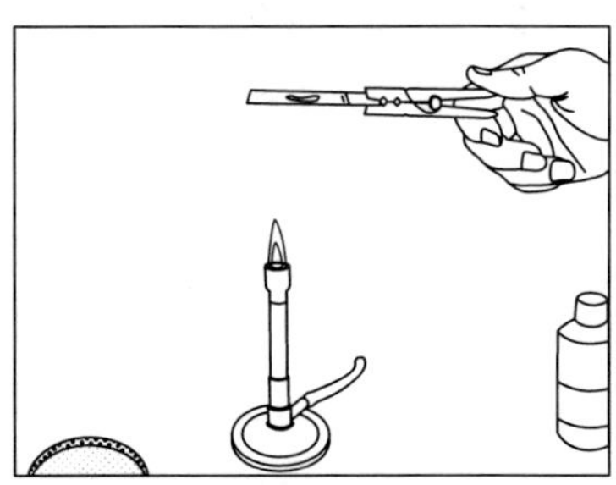

8. Heat fix the slide by passing it over the top of the flame a few times. When the slide is cool, stain the sample according to the following procedure.

Staining Procedure for Smear Preparations:

1. Place your heat-fixed slide over a staining rack and flood the smear with 2-3 drops of methylene blue dye.
2. Allow the slide to sit for 1 minute.
3. Rinse off the dye with distilled water.
4. Gently blot the slide with absorbent paper (do not rub the smear while blotting it) and let it air dry.
5. Examine under oil immersion. You do not have to use a coverslip for this preparation; the oil can be applied directly to the stained material.
6. Draw your observations in the appropriate section of the datasheet.

The shape of bacteria cells is a useful feature used in their identification (Fig. 2). The basic cell shapes include rods (as single cells, or linked together in chains, or curved—#4, 8, and 7); round-shaped cocci (as single cells, chains, or clusters—#1, 2, and 3); and spiral or helical cells called spirilli (#5).

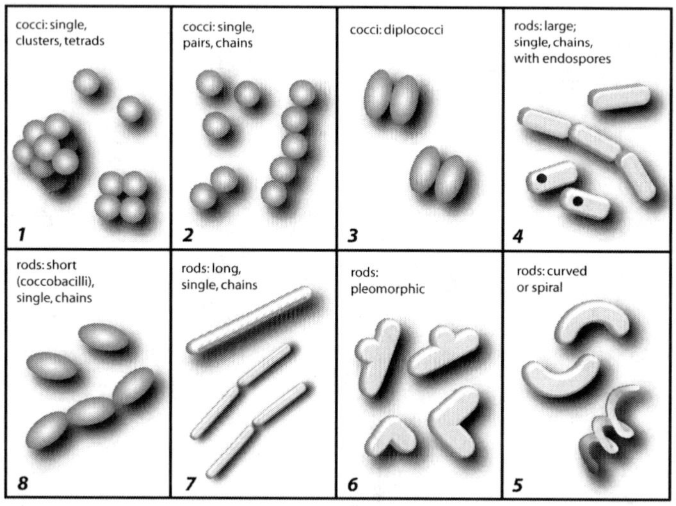

Figure 2. Examples of bacterial cell shapes and arrangements.

PROCEDURES (DAY 3)

Examination of Streak Plate:

In a dark room, observe the plate(s) that you streaked as before.

Prepare smears of any luminescent colonies, then stain them as described in the previous section. Observe under oil immersion, and describe the characteristics of those bacteria in your datasheet.

DVD:
The procedure for performing a Gram Stain is demonstrated in the video chapter for this exercise.

If your instructor wishes, you may perform the Gram Stain on some of the bacteria in your samples. This standard procedure in microbiology divides bacteria into two groups—Gram-negative or Gram-positive—depending on which color they appear after the staining process.

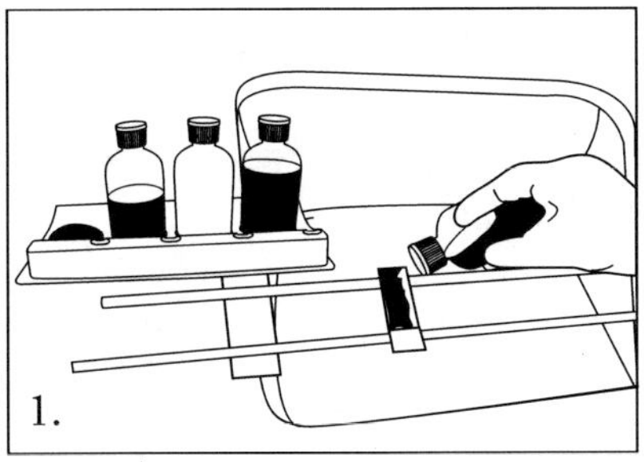

Crystal Violet - 1 minute

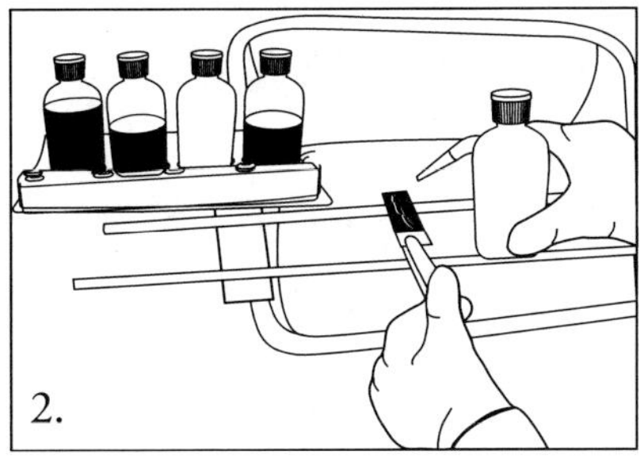

Rinse with tap water

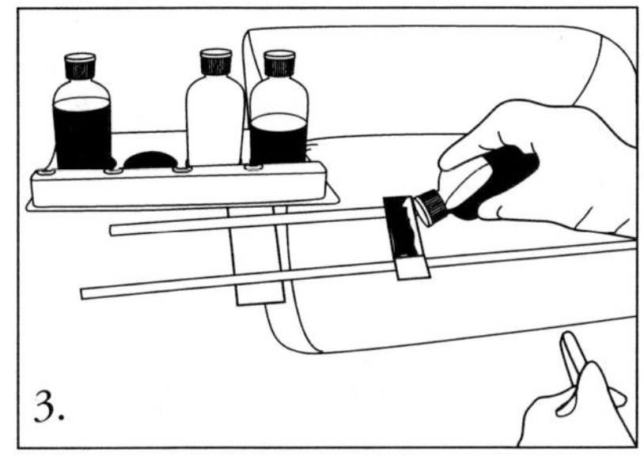

Iodine - 1 minute

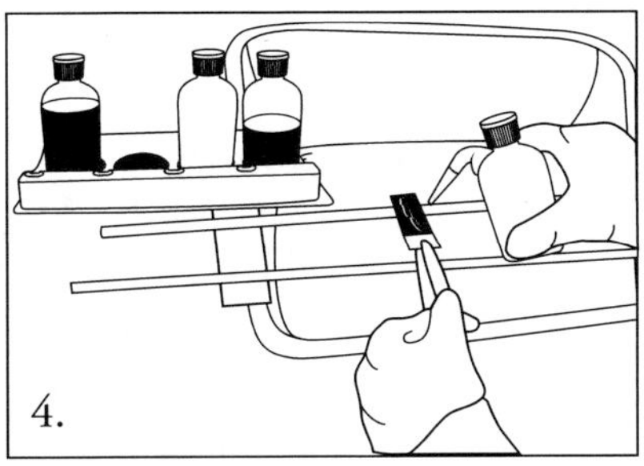

Rinse with tap water

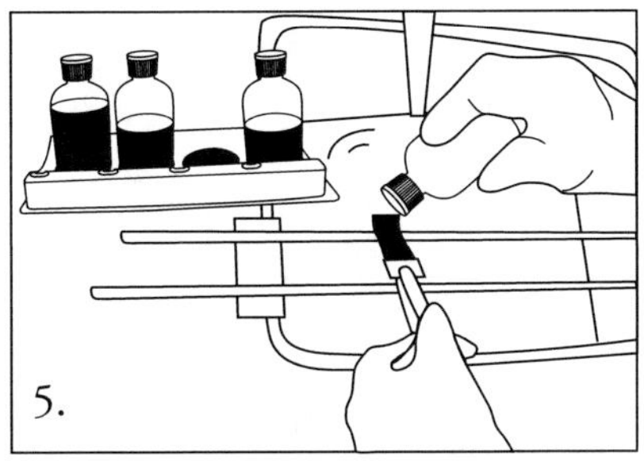

Alcohol drip - 30 seconds

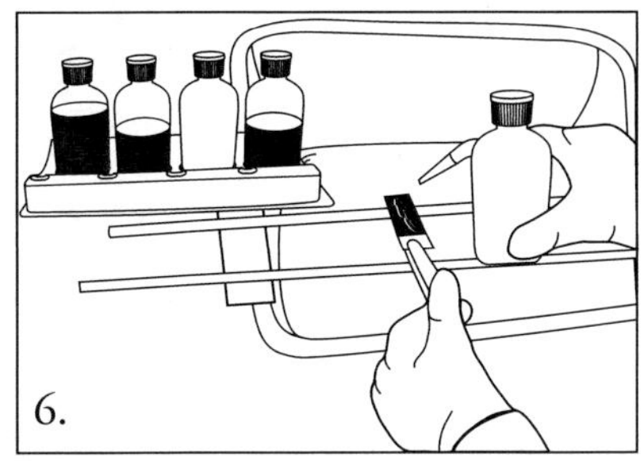

Rinse with tap water

Safranin - 1 minute

Rinse with tap water

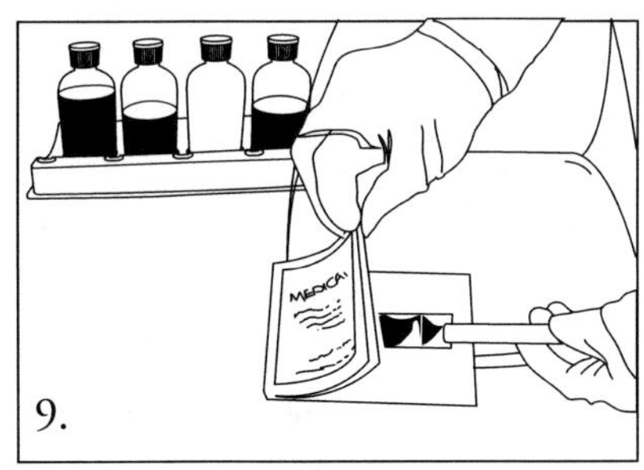

Blot dry

Performing a Gram Stain:

Prepare a heat-fixed smear preparation, as described in the previous steps.

1. Set the slide over a staining tray and add a few drops of crystal violet stain onto the smear. Wait 1 minute.
2. Hold the slide at an angle with a clothespin and gently rinse the dye off with tap water for a few seconds.
3. Flood the slide with a few drops of iodine solution and leave on for 1 minute.
4. Rinse the iodine off, as in Step 2.
5. Hold the slide at an angle and drop 95% ethyl alcohol onto the smear, one drop at a time for 30 seconds, or until no more purple stain rinses off.
6. Rinse with water, as above.
7. Add several drops of safranin stain and leave on for 1 minute.
8. Rinse with water, as above.
9. Gently blot the slide with absorbent paper (do not rub the smear while blotting it), and air dry.

10. Examine under oil immersion. You do not have to use a coverslip for this preparation; the oil can be applied directly to the stained material. Gram-negative bacteria appear purple, while Gram-positive bacteria appear pink or red. Identify each colony as Gram-negative or -positive.
11. Record your observations in the table of colony characteristics in the datasheet.

WEB EXPLORATIONS FOR FURTHER STUDY:

Explore these relevant websites to stimulate interest in marine microbiology.

Surfrider Foundation http://www.surfrider.org
Founded in 1984, this non-profit environmental organization is dedicated to the protection and conservation of the ocean and coastal regions across the world. The Blue Water Task Force program is devoted to water quality monitoring and educating the public about coastal water pollution issues. Coastal water samples are gathered through an extensive network of volunteers, and the water is analyzed for levels of ***coliform*** *bacteria (symbiotic intestinal species that indicate the presence of fecal material in coastal water).*

- **Activity**: Explore the homepage to learn about the monitoring programs currently being conducted by this organization. If you live in a coastal or Great Lakes state, click on the menu tab marked *Water Quality > State of the Beach* and on the next webpage that appears, choose your state from the drop-down menu to read about the health of your state's coastal waters. How many beaches are currently monitored in your state? How many days in the last year were beaches closed due to unsafe levels of fecal bacteria? What special programs or actions are being implemented in your state to improve coastal water quality?

U.S. Environmental Protective Agency BEACON
(Beach Advisory and Closing On-line Notification)
http://iaspub.epa.gov/waters10/beacon_national_page.main
This national database posts the occurrence of water pollution in coastal recreational waters. Every beach in the United States is displayed on a map linked to current and historical data on beach closures and warnings due to bacterial pollution.

- **Activity**: Visit the homepage and use the interactive map to find a list of all the monitored beaches within your county. Determine how many total days of beach closures have occurred this year within your county. How does this compare to last year or previous years? Discuss possible reasons why some beaches in your county are closed more frequently than others.

Name: ________________
Date: ________________
Section: ________________

LAB 5: MARINE MICROBIOLOGY DATASHEET

Results for Seawater Samples

Depth: ________m
Collection date: __________
Duration of incubation: __________

Collection location: ________________________
Incubation temp (°C): ________

R1: estimated # of colonies ______

R2: estimated # of colonies ______

Make a sketch of each plate in the circles above, or attach a digital photo of each plate to this datasheet. Number the colonies and use Figure 1 to help you describe the characteristics of each in the table below.

Colony #	Size (mm):	Color:	Shape (elevation)	Shape (margin)	Surface appearance	Gram stain (+/-)
1						
2						
3						
4						
5						
6						
7						

Results for Sediment Samples

Location of Sheltered Site: ________________ Depth: ________m Collection date: ___________
Location of Exposed Site: ________________ Depth: ________m
Duration of incubation: ___________ Incubation temp (°C): ________

Sheltered: Estimated # of colonies ______ Exposed: Estimated # of colonies ________

Make a sketch of each plate in the circles above, or attach a digital photo of each plate to this datasheet. Number the colonies and describe their characteristics as you did for the seawater samples.

Which location harbored the most bacteria? Was your hypothesis made in the previous laboratory period supported by your observations? If not, revise your hypothesis and propose some reasons that might explain the difference.

Colony #	Size (mm):	Color:	Shape (elevation)	Shape (margin)	Surface appearance	Gram stain (+/-)
1						
2						
3						
4						
5						
6						
7						

Make drawings of stained bacteria from several of the labeled **seawater** colonies, and refer to Figure 2 to identify their cell forms (**rods**, **cocci**, or **spirilli**).

Colony #: _______ Magnification: ______

Colony #: _______ Magnification: ______

Colony #: _______ Magnification: ______

Colony #: _______ Magnification: ______

Colony #: _______ Magnification: ______

Colony #: _______ Magnification: ______

Make drawings of stained bacteria from several of the labeled **sediment** colonies, and identify their cell forms (**rods**, **cocci**, or **spirilli**).

Colony #: _______ Magnification: ______

Colony #: _______ Magnification: ______

Colony #: _______ Magnification: ______

Colony #: _______ Magnification: ______

Colony #: _______ Magnification: ______

Colony #: _______ Magnification: ______

Results for Streak Plate/Isolation of Bioluminescent Colonies

Duration of incubation: __________ Incubation temp (°C): ________

Make a sketch in the circle above, or attach a digital photo of the streaked plate to this datasheet. Number several colonies and describe their characteristics as you did for previous samples.

Did you obtain a pure culture of luminescent bacteria (i.e., do all the colonies on your plate appear identical in appearance)?

If not, how many different types of bacteria do you see?

Colony #	Size:	Color:	Shape (elevation)	Shape (margin)	Surface appearance	Gram stain (+/-)
1						
2						
3						
4						
5						
6						
7						

Name: ____________________
Date: ____________________
Section: ____________________

WEB EXPLORATION REPORT

Website: ____

LABORATORY 6

Antimicrobial Effects of Marine Extracts

INTRODUCTION:

For thousands of years, human societies have used marine organisms to treat medical afflictions and disease. Today, as more terrestrial bacteria evolve resistance to our current arsenal of antibiotics, new and more effective treatments must be found. Therefore, pharmaceutical companies show considerable interest in commercially developing medicines and antibiotics derived from substances produced by marine organisms. Within the last decade, the rapidly growing field of **marine pharmacology** has exploded due to new analytical techniques developed in molecular biology, as well as new technologies in ocean exploration that have made it possible for marine biologists to collect organisms from deeper and more extreme habitats.

Recently, a study found that 95% of 54 species of Indo-Pacific seaweeds showed some antimicrobial activity against **pathogenic** bacteria: three-quarters of the seaweeds were active against two or more bacteria species, and 20% of the seaweeds tested had **broad-spectrum** effects against 4 bacteria species.[1] Similarly, on the Mediterranean and Atlantic coasts of Spain, 55 out of 82 seaweeds tested showed antimicrobial activity against one or more species of harmful microorganisms.[2]

In this exercise, you will investigate several seaweed species for potential antimicrobial effects by using a disk diffusion method. A paper disk is saturated with seaweed solution and placed on an agar dish culture of bacteria. Biochemicals produced by the seaweed will diffuse out from the disk into the surrounding agar, and any chemicals that either restrict bacterial growth or kill the cells directly will result in a **zone of inhibition** around the disc that will be noticeable in the following lab period. This same procedure is used by researchers at the Food and Drug Administration and other clinical laboratories to determine the effectiveness of various antibiotics on specific pathogens.

Materials:

- ❑ Petri dishes containing nutrient agar
- ❑ Paper punch
- ❑ Filter paper
- ❑ Forceps
- ❑ Bunsen burner (flame source), striker
- ❑ Sterile distilled water
- ❑ Adhesive tape
- ❑ Wax pencil/china marker
- ❑ Seaweed extracts labeled "Boiled" and "Ground up"
- ❑ Test tubes
- ❑ Disinfectant solution and paper towels
- ❑ Metric ruler

[1] Puglisi, M., Engel, S., Jensen, and W. Fenical. 2007. Antimicrobial activities of extracts from Indo-Pacific marine plants against marine pathogens and saprophytes. *Marine Biology,* Vol. 150(4): pp. 531-540.

[2] Salvador, N., Gomez-Garreta, A., Lavelli, L., and M. A. Ribera. 2007. Antimicrobial activity of Iberian macroalgae. *Scientia Marina* 71(1): pp. 101-114.

DAY 1 PROCEDURE:

1. Wash your hands thoroughly and clean the work surface of your lab bench using paper towels and disinfectant. Why is this first step always important whenever working with bacteria?

2. Using a paper punch, punch out several disks of filter paper into a sterile Petri dish. Do not touch the disks with your fingers. (Why?)
   Use forceps to transfer several of the disks into a small beaker of each of the different seaweed extracts provided by your instructor.
3. Use a sterile cotton swab to collect bacteria from various surfaces around the classroom: cell phones, floor, shoe, doorknob, lab sink, etc.
4. Place the collection swab into a test tube half full of sterile water, and shake the swab vigorously to dislodge bacteria into the water.
5. Withdraw the swab and press it against the inside of the test tube to squeeze out the excess liquid; then use the swab to completely streak a nutrient agar plate as follows.
6. Swab the whole surface of the plate, going in one direction; then turn the plate 1/3 turn and swab the entire surface again.
7. Turn the plate 1/3 turn, and repeat swabbing the plate for a third time. When you finish, put the swab back in the test tube. Cover the plate and let it stand for 5 minutes to allow the agar to absorb the liquid.
8. Sterilize the tips of a forceps by heating them in the flame of a Bunsen burner until the tips glow red; then remove and let them cool. Do not allow the forceps to touch anything before the next step. (Why?)
9. Use the forceps to pick up a disk from the seaweed extract, and allow the excess solution to drip off the disk for a few seconds. Remove the lid from the nutrient agar plate; then place the disk directly on the surface of the media, in the center of the plate. Gently press the disk onto the surface with forceps, but do not press the disc into the agar media or let the forceps touch the agar. Replace the cover on the nutrient agar plate. (Why is it important that the forceps not touch the agar?)
10. Using the same swab as before, repeat Steps 5–9 to prepare a replicate plate of the first plate.
11. Turn both of the plates upside down, secure their lids with tape, and label each plate with your name, the seaweed extract used, the replicate number, and the date. The plates will be incubated at 35°C until the next lab period.

You have now made 2 experimental plates, which are replicates of one **experimental treatment**. The experimental treatment consists of exposing bacteria in the media plates to a seaweed extract. A key part of many laboratory experiments is the presence of a **control**, which, in this case, is a third media plate that you will prepare next. The control plate differs in only one way from the experimental plates, in that the control lacks the **experimental variable**. All other variables that could conceivably affect the outcome of the experiment must be controlled (kept as similar as possible) between the experimental plates and the control plate. What is the experimental variable in the 2 plates that you just prepared?

If your intention is to determine if a seaweed extract has any effect on bacterial growth, how must you prepare your control plate? Think about this for a minute before you continue reading any further.

Discuss the following choices with your lab partner. Which would best describe the appropriate plate to use as the control in this experiment?

- A streaked Petri dish with no paper disk and no seaweed extract
- An unstreaked Petri dish with no disk
- An unstreaked dish with a paper disk soaked in seaweed extract
- A streaked disk soaked in sterile distilled water.

Describe what you have decided to prepare.

Check your answer with your instructor, and then prepare the control plate. Label it as "Control" and include your initials and the date.

> When you are done, wash your hands thoroughly, and disinfect your workspace according to instructions from your teacher.

DAY 2 RESULTS:

Examine the Petri dishes, but do not open them. Compare bacterial growth between the experimental and control plates. Bacterial growth may appear as shiny or cloudy material of various colors on the agar surface. If the seaweed extract showed antimicrobial activity, there will be a clear area around the paper disk, indicating that a substance diffused out from the paper into the agar and inhibited the growth of microbes or killed them. Use a plastic ruler to measure the diameter of the zone of inhibition (if present). Fill in the table and share your results with the class. At the conclusion of the exercise, discard the Petri dishes in the biohazard container or according to instructions from your teacher. Wash your hands thoroughly, and disinfect your workspace.

Treatment (Alga species)	Boiled sample Zone diameter (mm)	Ground-up sample Zone diameter (mm)
1.		
2.		
3.		
4.		
3% Hydrogen peroxide		
1% Bleach solution		

WEB EXPLORATION FOR FURTHER STUDY:

Center for Marine Biotechnology and Biomedicine
http://sio.ucsd.edu/About/Research_Overview/Marine_Biotechnology_and_Biomedicine/
This research division, housed at Scripps Institute of Oceanography, is the world's largest coordinated biomedical research program devoted to exploring and developing new medicines and technologies based on marine life.

- **Activity**: Perform a Web search to investigate different types of drugs derived from marine invertebrates. To start, use some or any of these search terms:

Ecteinascidin	marine pharmacology	marine microbiology
Topsentin	antitumor	marine medicine
Lasonolide	antiviral	marine drug discovery
Bryostatin	marine pharmaceuticals	
Pseudopterosins	marine biomedical research	

What species of marine organisms produce the chemicals from which the drugs listed above were derived? What do these particular animals have in common? Suggest possible reasons why these particular organisms produce such potent chemicals. What current research is being conducted at the Center for Marine Biotechnology and Biomedicine, and how will patients with cardiovascular disease, arthritis, and cancer benefit from this research?

Name: ________________
Date: ________________
Section: ________________

REVIEW:

1. Did any of the seaweeds show antimicrobial effects?
2. If so, how was the growth of bacteria and/or fungi affected?
3. Rank the species by their effectiveness, from highest to lowest.
4. Would you describe any of the seaweed extracts you tested as having **broad spectrum** or **narrow spectrum** antibiotic effects? Explain.
5. If any of the ground-up seaweeds showed antimicrobial properties, did boiling them reduce their effectiveness? Explain why this may be so.
6. What was the experimental variable in this investigation?
7. Name several aspects of this experiment that were replicated. Why is it important that replication occurs in a controlled study? How does replication affect your level of confidence in your experimental results?
8. What experimental conditions needed to be held constant for you to make a valid comparison of your results with those of your classmates?
9. If you were a biologist hired by a pharmaceutical company interested in developing new drugs, can you think of any marine species that may seem promising to investigate for potential antibiotic properties? (Hint: What kinds of marine organisms survive well in polluted waters with high bacteria concentrations?)

Name: ____________________
Date: ____________________
Section: ____________________

WEB EXPLORATION REPORT

Website: ____

LABORATORY 7

Observing Plankton

> *"It is true that the lack of fishing and dredging equipment deprived us of the excitement of bringing up fantastically shaped monsters from the deep. But in the plankton-nets we can catch a hundred bizarre forms to every one recovered from a dredge; we can find animals painted with the colors of the rainbow, whereas the deep-sea organisms are either black or red. Anyone who has once seen these exquisite creatures through a microscope will never again envy the person with a deep-sea dredge."*
>
> from *The Last Cruise of the Carnegie* (1932) by J. H. Paul.

INTRODUCTION:

Plankton is a collective term for aquatic organisms that drift throughout the water column and cannot swim against a current. The photosynthetic plankton is termed **phytoplankton,** and animal forms are the **zooplankton**. Planktonic organisms range in size from the **ultraplankton** (cyanobacteria; cells <0.005 mm in diameter) to **megaplankton** (animals larger than 2 mm). While generally thought of as small or microscopic, some plankton can reach massive sizes: certain jellyfish and colonial **salps** have bodies over 12 m long. Plankton is usually collected by towing a net. The plankton net retains drifting objects larger than the size of the mesh openings in the net. However, some of the larger, more mobile zooplankton species are able to avoid a slow-moving net, while delicate, gelatinous forms are usually destroyed using this method; thus, those types of organisms are underestimated in the sample. But with the development of remotely operated vehicles, robotically-controlled underwater sampling devices, and three-dimensional video surveys of the water column, marine biologists are finding new ways to collect and observe delicate members of the zooplankton community, and to discover new species from greater depths. In this laboratory, you will observe both preserved and live samples of coastal marine plankton.

DVD:
Watch the video chapter "Examples of Zooplankton" and be able to answer the questions posed on screen.

OBJECTIVES:

The goals of this lab are for you to:

- Distinguish between **autotrophic** and **heterotrophic** planktonic organisms.
- Recognize the most common representatives of phytoplankton and zooplankton found in samples from coastal regions.
- Distinguish zooplankton as either **holoplankton** or **meroplankton.**
- Understand the characteristic adaptations of planktonic organisms to their habitat.

EXERCISE I: PHYTOPLANKTON

The importance of phytoplankton to the functioning and maintenance of life on Earth cannot be overstated. Nearly half of the oxygen in our atmosphere is the result of the photosynthetic activity of these tiny drifting organisms, which also serve as the first link of the oceanic food chain. Without phytoplankton to contribute to the formation of the present atmosphere, terrestrial animals would never have evolved, and marine animal life would not be widely distributed throughout the oceans, as it is today. Two prominent groups of phytoplankton, the **diatoms** and **dinoflagellates**, can be found in any sample of seawater taken from any coastal location, and their ubiquity throughout the **photic zone** makes them key organisms to study and identify.

Diatoms and dinoflagellates may appear brownish under the microscope due to pigments they use in addition to chlorophyll for photosynthesis. When a dense concentration of dinoflagellates occurs in coastal waters, the water may become brownish or even reddish in color, hence the phenomenon called "**red tide.**" Some dinoflagellates produce toxins that kill fish, and many species are bioluminescent, producing a greenish glow when there is water motion, such as crashing waves or disturbance by a ship's wake.

1. First examine the prepared slides of diatoms at several magnifications. Your slide may contain numerous fragments of diatoms, but look for intact specimens. Diatoms are single-celled organisms living singly as individuals or linked together in chains. Each cell is covered with a thin glass shell made of two lids (**valves**) joined together at a **girdle band** (Fig. 1, A–D). The valves may have numerous spines, bumps, or other ornaments on their surface, but notice how all are perforated with numerous pores. Why do you think all diatoms have these pores? Answer the following questions to guide your thinking.

What critical biological process occurs across the membranes of all cells?

Could this process occur in a cell encased within a nonporous glass shell?

So what, then, is a function of these pores?

Recognize the difference in shape between **pennate** and **centric** diatoms. On pennate diatoms, identify the **raphe** (Fig.1b). Pennate diatoms are benthic and crawl across surfaces, and the raphe is in contact with the surface. Centric diatoms are planktonic and do not have a raphe. Therefore, what is the function of the raphe?

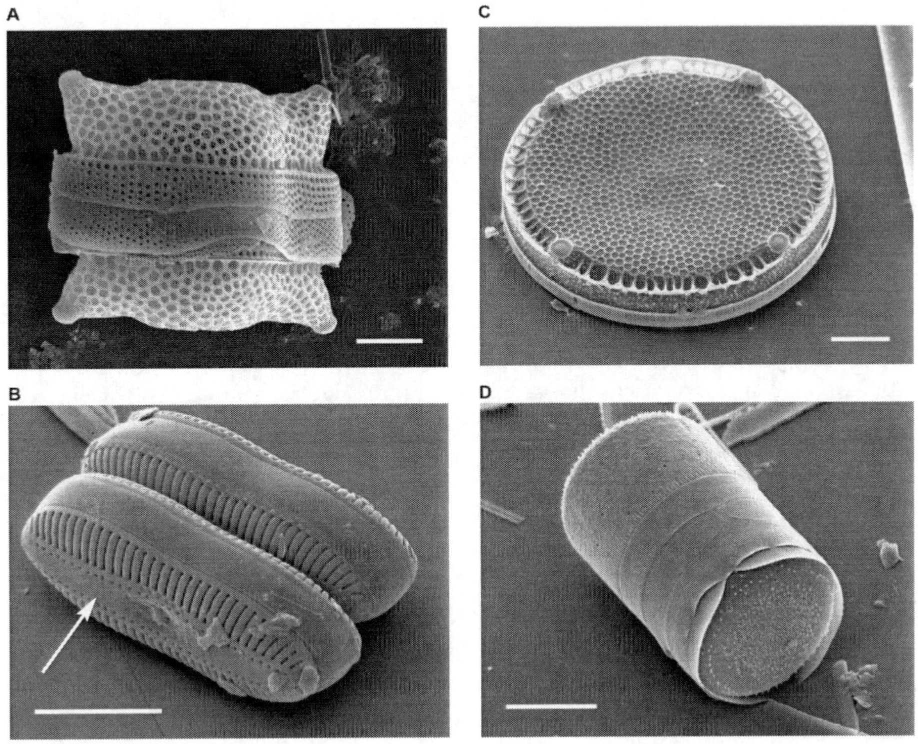

Courtesy of Mary Ann Tiffany

Figure 1. Scanning electron micrographs of diatoms. a) *Biddulphia reticulata*, a centric diatom with a wide girdle band (size bar=10µm). b) *Diploneis* sp. 2 pennate cells. Arrow points to raphe (size bar=10µm). c) *Eupodiscus radiatus*, a centric diatom (size bar=20µm). d) *Melosira varians*, a centric diatom (size bar=10µm).

2. Make sketches of centric, pennate, and chain forms in the boxes below.

Centrics:	Pennates:	Chains:

3. Dinoflagellates exhibit a variety of shapes, and the kinds most easily identified are those in the genus *Ceratium*. Species resemble three-pointed anchors, some with their points curved upward **(Fig. 2)**. Find several members of this genus on your slide, and sketch one in a box on the following page.

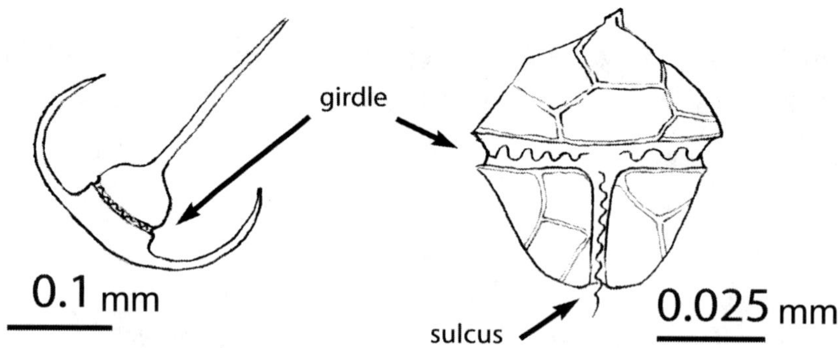

Figure 2. Dinoflagellates are unicells that move through the water with **flagella** (usually 2, hence the name). The flagella of most species are arranged within a circular **girdle** and a vertical **sulcus**.

Next, examine some living phytoplankton.

> NOTE: Seawater is very corrosive to instruments, so MAKE SURE you do not get seawater on the microscope (especially the lenses). If you do, wipe it off immediately. At the conclusion of the lab, wipe the stage with a paper towel moistened with fresh water, then with a dry paper towel.

- Using a pipette, place a few drops of sample onto a slide and cover with a coverslip.
- Place the slide onto the compound microscope and scan at the lowest power to locate clumps of diatoms. To see the plankton better, you will have to play with the aperture to reduce the light coming up through the sample to create some contrast.
- Examine the living cells. Can you see chloroplasts? The brown color of the cell contents is due to golden pigments and, in some cases, oil droplets (diatoms make oil for food storage, and this low-density material also lowers the overall density of the cell, helping to increase the diatom's buoyancy).
- Are any of the pennate diatoms crawling? You might also see some of the colonial diatoms moving in a gliding motion (look for long chains that are expanding and contracting like an accordion).
- Use the illustrations and local identification guides (if available) to identify both solitary and colonial forms of centric and pennate diatoms. Draw and label at least 3 different types in the following boxes.

Common name: Genus:	Common name: Genus:	Common name: Genus:

Common name: Genus:	Common name: Genus:	Common name: Genus:

EXERCISE II. ZOOPLANKTON

Zooplankton consists of two major types: **holoplankton** (animals that remain planktonic throughout their entire life cycle), and **meroplankton** (those that spend the early portion of their lifespan drifting in the water column before developing into **nektonic** or **benthic** adults). Pour some of the zooplankton sample into a Petri dish and look at it under the dissecting microscope. Remove individual specimens with a pipette and place them into a depression slide, cover with a coverslip, and examine under the compound microscope. If you have living specimens, it may be necessary to slow down the movement of those plankters with methyl cellulose if they are swimming too fast to observe.

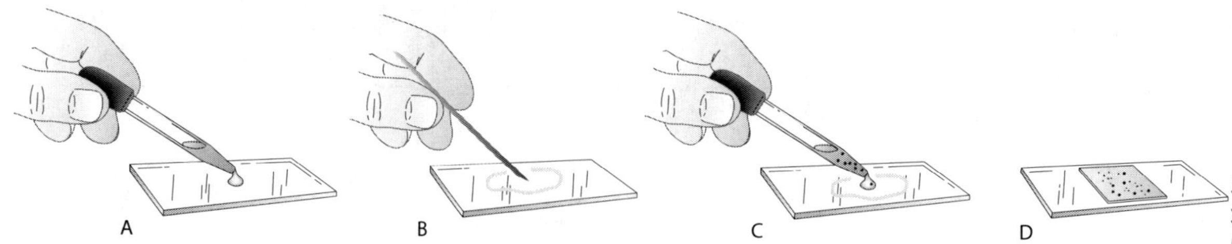

A. Add a drop of methyl cellulose to a clean slide
B. Use a toothpick to make a circle of liquid slightly smaller than the size of a coverslip
C. Add 1-2 drops of plankton into the center of the circle
D. Add a coverslip and examine under the compound microscope

Fig. 3 Procedure for making a wet mount with methyl cellulose for viewing highly mobile organisms. As the methyl cellulose diffuses into the plankton sample, it thickens the seawater, reducing the speed of swimming animals.

Use the illustrations in the following section to identify the most common types of animal plankton in your sample. Plankton can be quite diverse in terms of the number of taxonomic groups represented. There is no telling what you might have, as the composition of zooplankton is influenced by the season, the ocean currents, the water temperature, and a host of other biological and environmental factors that vary geographically and temporally. However, coastal samples usually include a few dominant groups, namely herbivorous copepods (which will be by far the most abundant type of zooplankton in your sample), and the larvae of barnacles, annelids, and molluscs. For the purposes of this exercise, try to identify the organisms to phylum or class; identification to lower taxonomic levels requires much more time and expertise.

Draw the most abundant types of zooplankton in the boxes on the following pages. For each type of zooplanktonic organism you observe, follow these guidelines:

- Describe any color(s) (patterns, locations on the body of pigmented areas).
- Label unique anatomical structures, and speculate on their purpose/adaptive value.
- Describe movement. Is it smooth and gliding? This motion is usually due to the action of cilia. If movement is jerky, are there beating appendages, or does the animal contract its body?
- Describe any behaviors you observe (are animals creating feeding currents, engaging in predation or reproductive activities?)

Now, add a tiny speck of dilute Carmine powder solution to your sample, using the tip of a toothpick. Look for currents of water circulating around the anterior or head region of any zooplankton. These currents are especially noticeable around a copepod. For what reason(s) do these animals circulate water in this way?

When you are finished observing the organisms on the slide, remove the coverslip and use a squeeze bottle of seawater to rinse the plankton into a small beaker. Wipe off the slide with a paper towel; then make a new slide using fresh plankton from the Petri dish.

Phylum: Common name:	Phylum: Common name:	Phylum: Common name:

Phylum: Common name:	Phylum: Common name:	Phylum: Common name:

Phylum: Common name:	Phylum: Common name:	Phylum: Common name:

Phylum: Common name:	Phylum: Common name:	Phylum: Common name:

Phylum: Common name:	Phylum: Common name:	Phylum: Common name:

Phylum: Common name:	Phylum: Common name:	Phylum: Common name:

QUANTIFICATION OF A PLANKTON SAMPLE

When collecting plankton, it is important for biologists to record other information that helps in the understanding of plankton ecology and seasonal abundance. If any of this information has been collected along with your sample, record it below.

Location of tow: ______________________	Beaufort Sea State: ________________
Plankton net mesh size or number: _________	Date of tow: ______________________
Net diameter: _________________________	Water temperature: _________________
Towing depth (m): _____________________	Salinity (ppt): _____________________
Tow start time: _______________________	Water color: _______________________
Tow ending time: ______________________	Type of tow (vertical, horizontal): ________
Duration of tow (as a fraction of an hour): ____	Light penetration (Secchi depth): ________
Vessel speed**: _______________________	Dissolved Oxygen (mg/l): _____________

**If speed was measured in knots (nautical miles per hour), then multiply the speed in knots by 1,853 to get the speed in meters per hour.

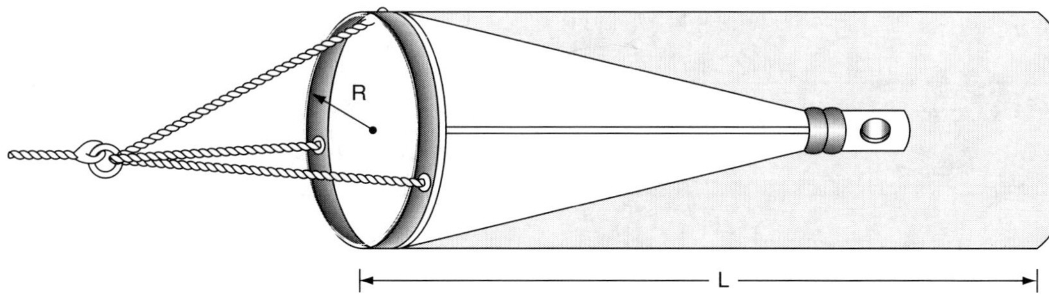

Fig. 5 The amount of water passing through a plankton net can be computed as the volume of a cylinder with length L and radius R.

The information above is used to determine how much water passed through the plankton net during the tow, and then to compute the density of zooplankton per standard volume of seawater (usually one cubic meter, or m^3).

Field data computation:

To compute water volume that passes through a net towed behind a boat:

$$m^3 = \pi * r^2 * L \qquad (\pi = 3.14159...)$$

m^3 = volume of water in cubic meters

volume = Area x Length (L = length of tow in m)

Area of circle = π* (radius of net in m)^2 (radius = 1/2 diameter of net)

L = (vessel speed) * (how long tow was conducted, in hours)

To compute the density of plankton organisms, pour the entire plankton sample from the collecting bottle into a graduated cylinder. After 10-15 minutes, the plankton will have settled to the bottom. Record the volume of plankton organisms in milliliters, then divide plankton volume (in ml) by water volume (in m^3) to obtain a standard density (ml/m^3).

Example: If volume of tow was 47.1 m^3, and the total volume of organisms collected was 3.5 ml, then density per m^3 = 3.5 ml / 47.1 m^3 = 0.074 ml/m^3.

Record your plankton density: __________ml/m^3

Plankton density provides a measure of how much food is available to other organisms higher up the food chain. Oceanographers compare this standardized density over time to discriminate monthly and yearly patterns in plankton abundance.

At the conclusion of this activity, wipe off all traces of seawater from the microscopes and any metal equipment that has come in contact with seawater.

Plate 1. SINGLE-CELLED PLANKTONIC PROTISTS

The foraminiferans and radiolarians are two groups of shelled, unicellular predators in the world ocean. The shells of forams are made of calcium carbonate, while radiolarians have skeletons of glass. They feed by sending out thin extensions of their cell contents (called **pseudopods**) through tiny holes in their skeletons. Organic material and other planktonic organisms stick to the pseudopods and are ingested by the cell. When these tiny predators die, their skeletons sink to the bottom, forming ocean sediments up to several kilometers thick. Diatoms also contribute to ocean sediments when they die and sink, but there are no dinoflagellate remains in sediments; their cell covering is made of biodegradable cellulose.

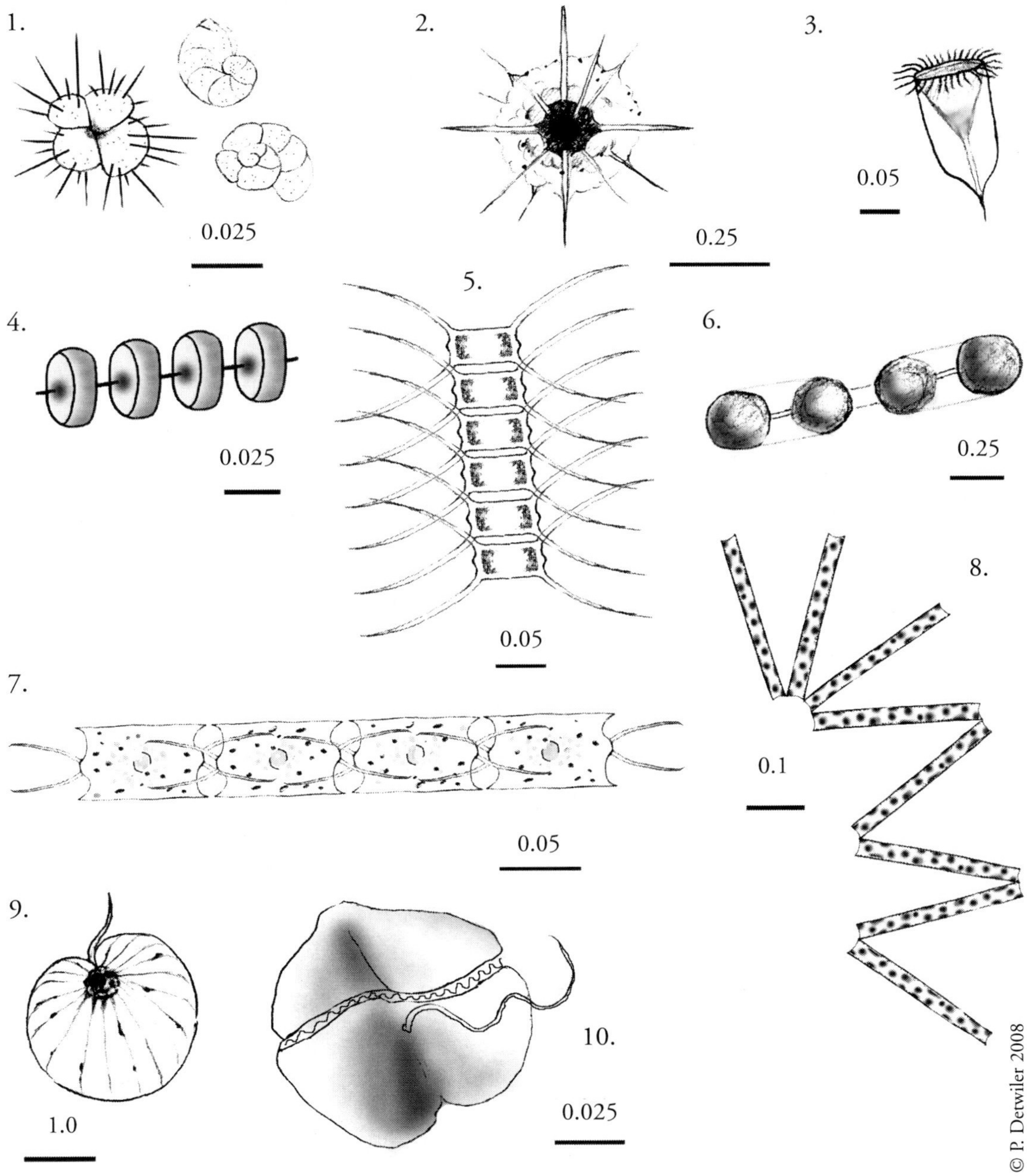

1. Forameniferan *Globerigina*
2. Radiolarian
3. Tintinnid
4. Diatom chain *Thalassiosira*
5. Diatom chain *Chaetoceros*
6. Diatom chain *Skeletonema*
7. Diatom chain *Biddulphia*
8. Diatom chain *Thalassionema*
9. Dinoflagellate *Noctiluca scintillans*
10. Dinoflagellate *Karenia brevis*

Scale bars for all figures are in mm.

Plate 2: PHYLUM ARTHROPODA: Class CRUSTACEA

This is the largest phylum in the animal kingdom, meaning that there are more species of arthropods than any other type of animal on Earth. **Crustaceans** are the most abundant type of zooplankton in ocean water samples, and, of these, **copepods** are the dominant group. Inspect the following pictures of different planktonic crustaceans. What is the key feature of all arthropods that you see in these images? Can you recognize two common adaptations for a planktonic lifestyle?

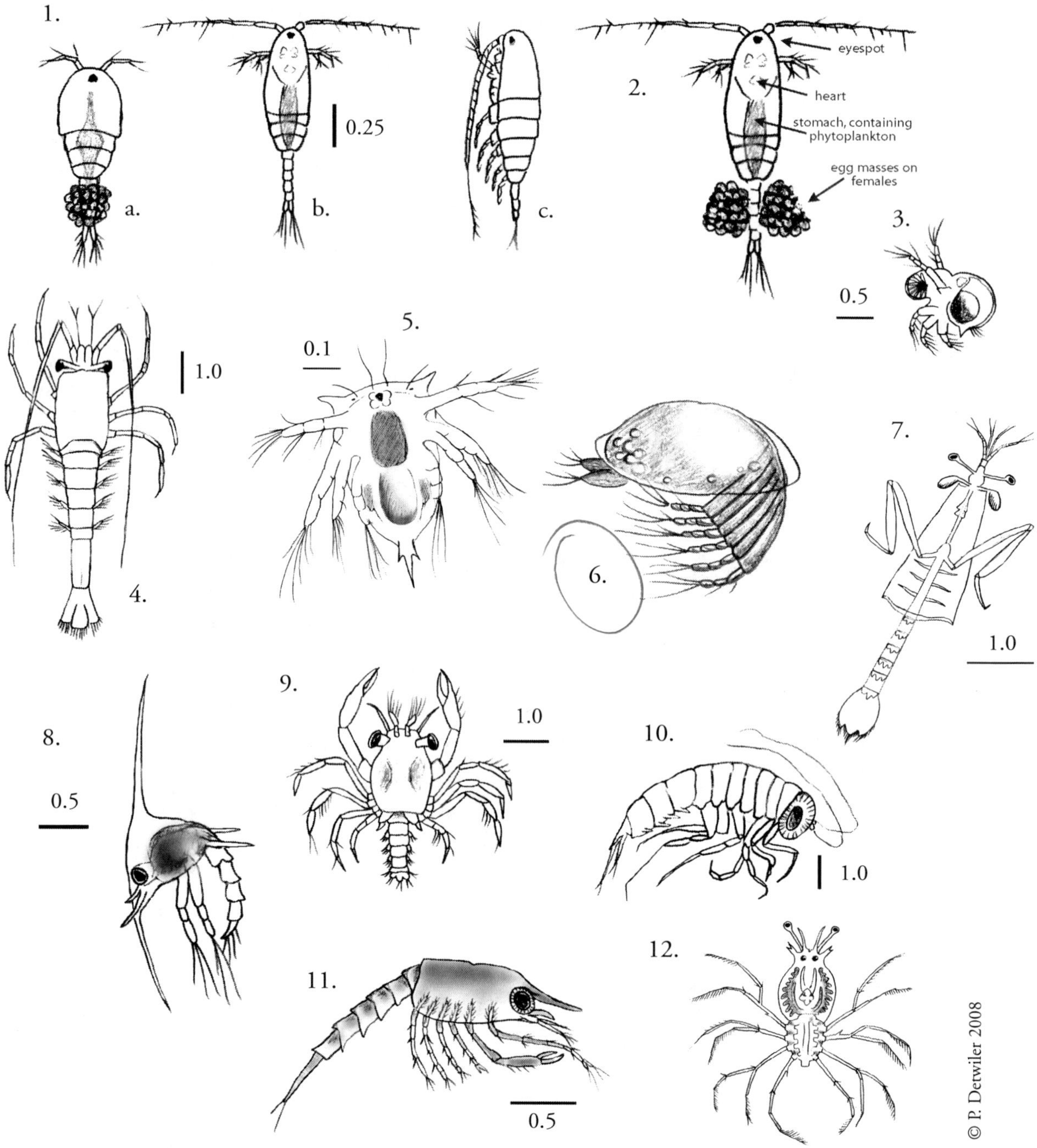

1. Copepods–a: cyclopoid; b: calanoid; c: side view
2. Copepod anatomy
3. Cladoceran
4. Zoea larva of shrimp
5. Nauplius larva of barnacle
6. Cyprid larva of barnacle
7. Larva of mantis shrimp
8. Zoea larva of crab
9. Megalops larva of crab
10. Amphipod
11. Mysis of lobster
12. Phyllosome larva of lobster

Scale bars for all figures are in mm.

Plate 3: HOLO- and MEROPLANKTONIC ZOOPLANKTON: PROTOSTOMES

These phyla show a remarkable diversity in body forms, but these animals all begin their development with the mouth as the first opening that forms in the embryo. Many of these larvae have tufts or fringes of cilia to propel them through the water. Meroplanktonic species live in the plankton and grow until they become too big and heavy to float, at which point they sink to the bottom and metamorphose into their benthic form. Which 4 animals on this page are holoplanktonic?

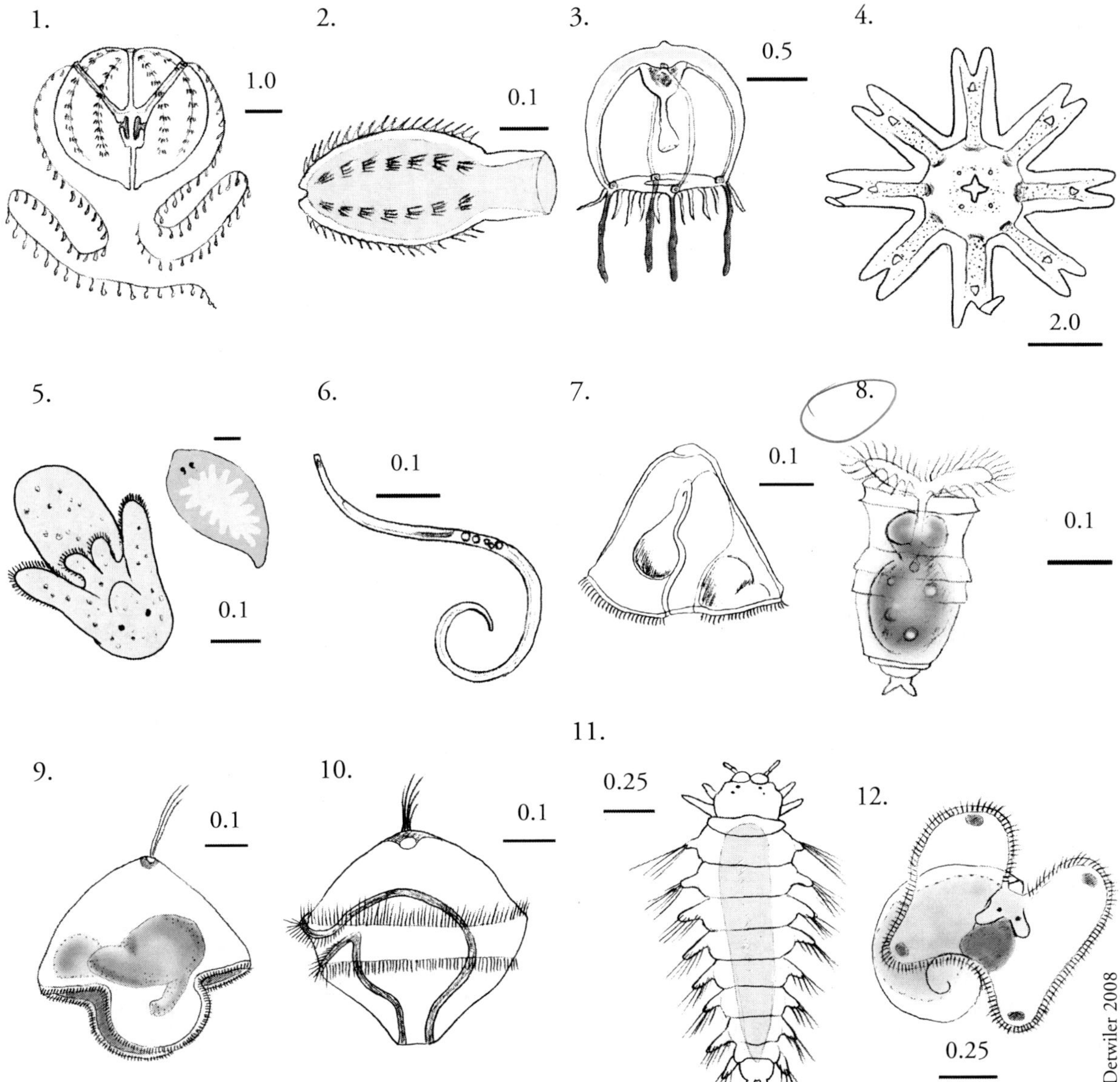

1. Phylum Ctenophora (Comb Jellies). Sea Gooseberry.
2. Phylum Ctenophora. Cydippid larva of a Comb Jelly.
3. Phylum Cnidaria. Medusa.
4. Phylum Cnidaria. Ephyra larva of jellyfish *Aurelia aurita.*
5. Phylum Platyhelminthes (Flatworms). Müller's larvae.
6. Phylum Nematoda (Roundworms).
7. Phylum Ectoprocta. Cyphonautes larva of a bryozoan.
8. Phylum Rotifera.
9. Phylum Nemertea (Ribbon worms). Pilidium larva.
10. Trochophore larva of Annelida and Mollusca.
11. Phylum Annelida. Polychaete larva.
12. Phylum Mollusca. Veliger larva of gastropod snail.

Scale bars for all figures are in mm.

Plate 4: HOLO- and MEROPLANKTONIC ZOOPLANKTON: DEUTEROSTOMES

In these phyla, the first opening that appears on the embryo becomes the anus. Transparent arrow worms are voracious predators that catch their prey with rows of hooked spines surrounding the mouth. Notice the cilia-lined bands fringing the lobes and arms of the various echinoderm larval stages, which both propel the animal and create currents to direct phytoplankton toward the mouth. The phylum Chordata is represented by eggs and fish larvae, and coastal plankton hauls often contain members of the subphylum Urochordata–the larval forms of benthic sea squirts (tunicates), as well as holoplanktonic larvaceans.

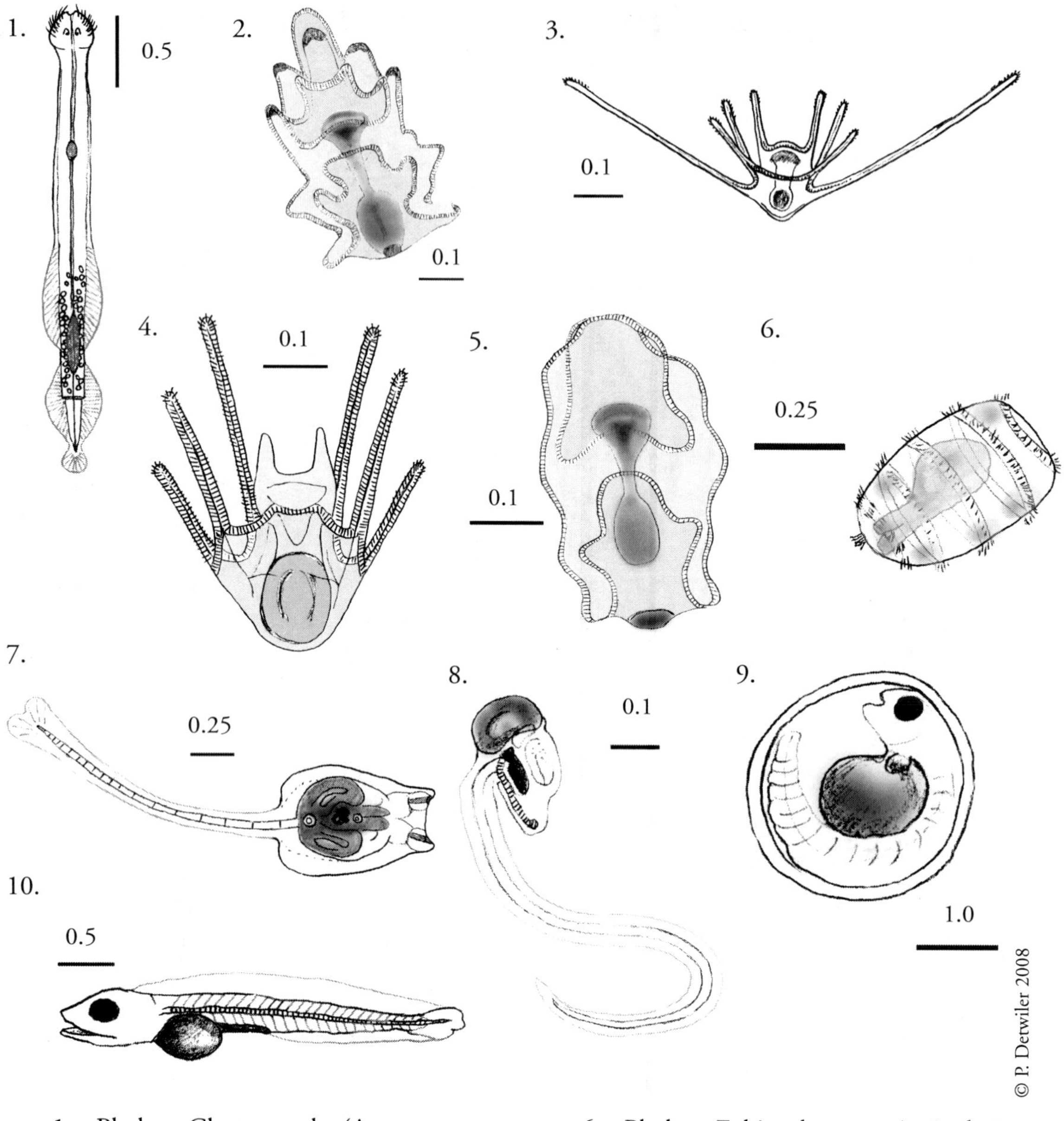

1. Phylum Chaetognatha (Arrow worms).
2. Phylum Echinodermata. Brachiolaria larva of a starfish.
3. Phylum Echinodermata. Ophiopluteus larva of a brittlestar.
4. Phylum Echinodermata. Echinopluteus larva of a sea urchin.
5. Phylum Echinodermata. Doliolaria larva of a sea cucumber.
6. Phylum Echinodermata. Auricularia larva of a sea cucumber.
7. Phylum Chordata. Tadpole larvae of a tunicate.
8. Phylum Chordata. The larvacean *Oikopluera*.
9. Phylum Chordata. Fish egg.
10. Phylum Chordata. Fish larva.

Scale bars for all figures are in mm.

LABORATORY 8

The Effects of Pollution on Phytoplankton Growth

Phytoplankton are critically important to the global ocean ecosystem. Without them, higher forms of marine life could not exist. Because phytoplankton form the base of the oceanic food chain, changes in their abundance and diversity affect other species that depend on them for food. If environmental conditions are favorable, phytoplankton reproduce quickly, increasing their abundance in the water column. Consequently, populations of herbivorous **zooplankton** (which consume phytoplankton) also increase, followed by carnivorous zooplankton that feed on the herbivores. If an environmental stress, such as chemical pollution, causes phytoplankton numbers to decrease, the whole food chain may be negatively impacted.

Many types of human activities create pollution that interferes with normal phytoplankton growth. **Positive effector pollution** causes the number of phytoplankton to increase, which may be dangerous in some cases. For example, on the New England coast, agricultural runoff containing fertilizer stimulates the growth of the dinoflagellate *Gonyaulax*, causing a **bloom** (a high rate of algal reproduction within a short time period). This can turn the water reddish in a phenomenon formerly known as **red tide** (scientists now prefer the term **harmful algal bloom** (HAB) because not all dinoflagellate blooms turn the water red). *Gonyaulax* produces **saxitoxin**, a toxic chemical that becomes concentrated in the flesh of filter feeders like bivalves and in herbivorous zooplankton. Zooplanktivorous fish further concentrate saxitoxin, and if higher vertebrates (such as dolphins, seabirds, and humans) ingest contaminated shellfish or fish, a potentially fatal syndrome called **paralytic shellfish poisoning** may result.

Another type of pollution that destroys phytoplanktonic organisms is **negative effector pollution**. In this case, the rest of the food chain organisms depending upon the phytoplankton also decrease in abundance. Herbicide runoff from agriculture and residential home landscaping is under this category.

Pharmaceutical and personal care products (PPCPs) such as prescription drugs, soaps, shampoos, and other household cleaning and beauty products are appearing in many of the nation's waterways and water bodies that serve as sources of drinking water. Many types of prescription drugs are excreted in urine, but large amounts also enter the sewage system due to the widespread practice of flushing unused or expired medications down the toilet for disposal. Many of these chemicals are not broken down but pass through water treatment plants intact and persist in the environment. Some of these may **bioaccumulate** within the aquatic food chain. Others become more toxic when they break down into other chemicals and may act as **endocrine disruptors**. The long-term effects of exposure to these chemicals for humans are unknown. When PPCPs enter the marine environment, could they harm marine life? In this activity, you will make a hypothesis about the effects of one PPCP on the growth of a single species of marine life that represents phytoplanktonic organisms at the base of the food chain. You may then generate hypotheses about any potential effects this PPCP may have on the larger marine ecosystem.

The goals of this laboratory are for you to accomplish the following:

- Understand the impacts of pharmaceuticals and personal care products in aquatic environments.
- Demonstrate proper use of the spectrophotometer and proper technique for making a serial dilution of a chemical.
- Experimentally determine the effect of a household personal care product on the growth of a unicellular green alga.
- Consider how your use of personal care products may affect the environment.

BACKGROUND:

Dunaliella salina is a species of unicellular marine phytoplankton commonly found in highly saline water bodies (salinity >65 ppt). When its population density is high (when there are many cells per ml of water), the water may appear brown, red, orange, or yellow due to the abundant **carotenoid** pigments used as **accessory pigments** for photosynthesis.

Theory of Monitoring Growth:

Population growth rate is defined as the number of individuals added to the existing population within a given time period. In order to measure algal population growth, you need a mechanism to measure how the cell density changes over time. However, counting the number of cells added to the population is not practical. Because your organism is a green alga (and contains chlorophyll *a*), a more reasonable method is to measure the change in the amount of chlorophyll *a* in the sample over time. The more chlorophyll *a* that is present, the more cells there are in the sample. What kind of functional relationship exists between the number of algal cells in your sample and the amount of chlorophyll in your sample? Circle one: Positive (direct) / Negative (inverse)

You will take samples from pure cultures of *Dunaliella salina* and from those exposed to pollutants. Then you will use a machine called a **spectrophotometer** to monitor the absorbance of light at the wavelength of 670 nm, the absorbance peak for the chlorophyll *a* pigment. A spectrophotometer allows you to select a particular wavelength of light and shine it through a sample of liquid. A detector on the other side of the machine records the amount of light passing through the sample. This is termed the **transmittance** and is expressed as a percentage. The ratio between the starting amount of light and the amount of light the detector receives is the percentage of light transmitted through the sample.

You will convert readings of transmittance into absorption, using the following formula:

$$\text{Per cent absorption} = 1 - \text{transmittance.}$$

Your readings will give you an idea of how much *Dunaliella* is in the sample. For example, in your experiment, if you have X amount of the alga in a sample and the reading for transmittance is Y, then twice the amount of that organism (i.e., 2X) will give a reading of 1/2Y on the spectrophotometer.

To begin this investigation, bring to the laboratory a small amount of a PPCP that you use regularly at home. Choose any type of household chemical that ends up going down the drain or the toilet. This substance will be considered the **experimental variable**. You will dilute your chemical to a concentration of 1 part per million (1 ppm), then add it to *Dunaliella* cultures. At least twice a week, you will measure light absorption in the control and experimental groups to monitor cell population growth.

State your testable hypothesis below. Indicate your experimental variable and what effect you think it may have on the system being studied.

MATERIALS:

Assemble the following at your workstation:

- 200-ml beaker of Enriched Sea Water and 100-ml graduated cylinder
- 2 250-ml flasks with labeling tape and a permanent marker
- 50-ml beaker of *Dunaliella* culture solution with transfer pipette and 10-ml graduated cylinder
- Test tube rack with 4 test tubes
- 50-ml beaker of distilled water with pipette and 10-ml graduated cylinder
- Parafilm® and scissors

GENERAL PROCEDURE:

1. Prepare a diluted solution of your chemical pollutant, as explained below.
2. Prepare your control and experimental flasks and add the treatment (your PPCP) to the experimental flask.
3. Use the spectrophotometer to record light transmittance of samples taken from both flasks and record your data in Table 1.

Procedure for Serial Dilution of PPCP chemical:

1. Number 4 test tubes 1 through 4, and place them into a rack.
2. Fill each test tube with 9 ml of distilled water.
3. Add 1 ml of your pollutant to test tube 1 and mix thoroughly. You now have made a 10% solution of your chemical.
4. Remove 1 ml of the solution from test tube 1 and add it to test tube 2. Mix completely. This is now a 1% solution of your chemical.
5. Remove 1 ml from test tube 2 and add it to test tube 3; mix.
6. Remove 1 ml from test tube 3 and add it to test tube 4; mix.
7. After you have taken a spectrophotometer reading of the Experimental flask (see Steps 1-8 of the next procedure), add 1 ml from test tube 4 to the 100-ml algal solution in the Experimental flask. This flask now contains your chemical pollutant at a concentration of 1 part per million (1 ppm).

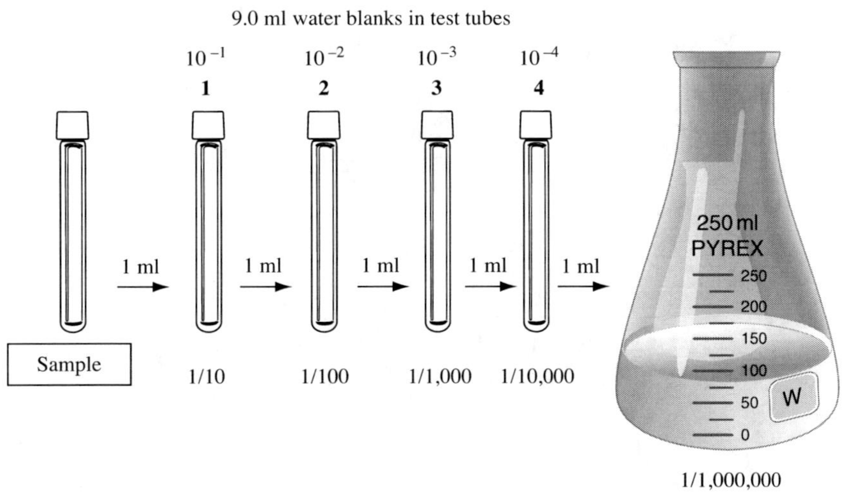

Figure 1. Serial dilution of a chemical pollutant to create a final concentration of 1 ppm.

Procedure for Preparing Samples and Using the Spectrophotometer:

1. Fill two 250-ml flasks with 90 ml of sterile Enriched Seawater solution.
2. Label both flasks with your initials, section, and date. Label one flask "Control" and the other "Experimental," with the name of your chemical pollutant.
3. Use a pipette to transfer 10 ml of *Dunaliella* stock solution into each flask.
4. Near the spectrophotometer is a cuvette (or small test tube) filled two-thirds full with Enriched Seawater solution. This is the **blank**, and it is used to set the "zero point" on the spectrophotometer.
5. ***Wipe off any moisture or fingerprints on the outside of the blank before inserting it into the Spectrophotometer.*** Moisture will damage the inside of the machine, and fingerprints will affect light transmission through the sample.
6. Refer to Figure 2. Turn the wavelength control knob to 670 nm (this is the wavelength of light that will pass through the sample). Set the scale to 0% absorbance (the display should read 0.000) for absorbance mode and 100% for transmission mode, using the zero control knob.
7. Insert the blank into the sample chamber, close the lid, and adjust the transmission to 100%.
8. Remove the blank and place it in its storage area. Do not discard the blank; you will use it for subsequent measurements on future days. Record transmittance of *Dunaliella* from each flask as follows:
   a. Gently swirl each flask before pouring a sample into a cuvette or tube.
   b. Dry the outside of the tube, place it in the compartment, and close the lid.
   c. After the display stabilizes, record % transmittance in the Data Table.
9. Now add 1.0 ml of your diluted chemical pollutant from test tube 4 into the Experimental flask, swirl to mix, and record transmittance.
10. Pour each sample down the drain after you take a reading. ***DO NOT pour the sample back into its flask.***
11. Rinse your sample tubes cuvettes with tap water and place them upside down to dry in the designated area.
12. Cover the tops of the flasks with Parafilm®, and place your flasks in the culture area. Record the environmental conditions (temperature, light level, and duration) in the Table 1.

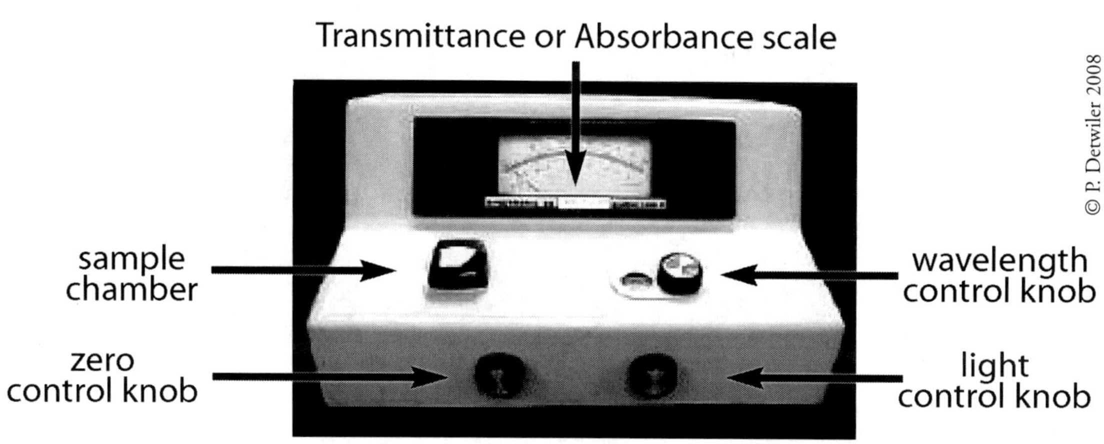

Figure 2. Parts of the spectrophotometer. This is a generalized illustration; the model you will use may differ. Your instructor will provide specific operating instructions for its use.

You will continue to collect spectrophotometric data over several days, as directed by your instructor. Over the next several days, *Dunaliella* may grow rapidly, forming an algal film on the bottom and sides of the flask. Before you take a reading, remove this film by swirling or shaking the flask steadily until all of the cells adhering to the sides of the flask are resuspended in solution.

Record your data in the following table:

Chemical name: ______________________________

Table 1. Spectrophotometer readings showing the response of *Dunaliella* to PPCP exposure.

	Experimental		Control		
Date:	% Transmittance (T) at 670 nm:	% Absorption (= 1 – T)	% Transmittance (T) at 670 nm:	% Absorption (= 1 – T)	Notes/Observations:

At the conclusion of the data-collection period, use spreadsheet software to create a graph depicting the change in light absorption by chlorophyll *a* (y-axis) over time (x-axis) in *Dunaliella* in the experimental and control flasks, as shown in Figure 3.

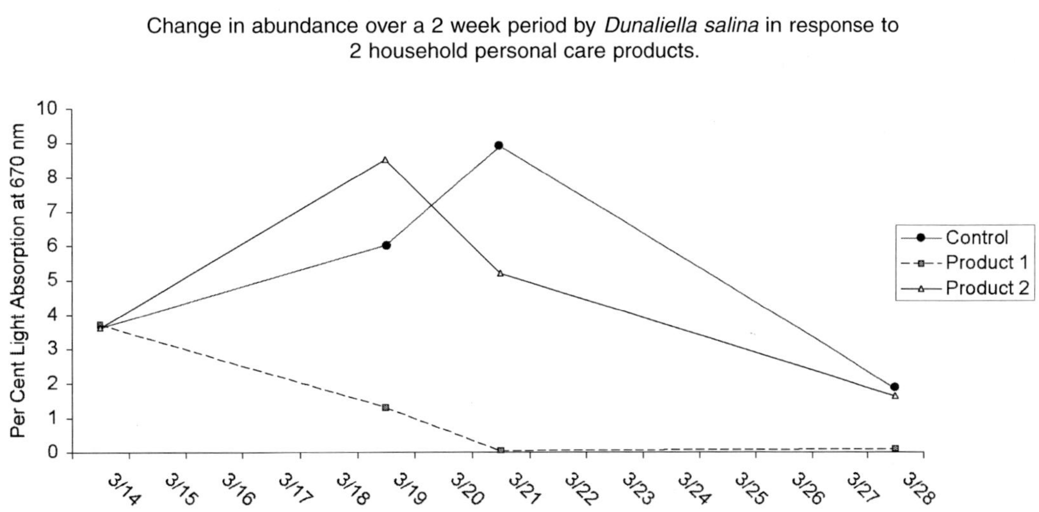

Figure 3. Example of proper format showing experimental results over a two-week period. Which of the two products is a positive effector pollutant? Which is a negative effector?

CONCLUSIONS AND QUESTIONS:

Name: ____________________
Date: ____________________
Section: ____________________

1. Was your hypothesis supported by your results? Why or why not?

2. Was your chemical a positive or negative effector pollutant for *Dunaliella*?

3. How do you think your chemical affected *Dunaliella*? Research the principle chemicals present in your PPCP to learn possible mechanisms for how they affect cells.

4. Provide 2 examples of sources of error in this experiment, and explain how each could have affected your results. If you did not swirl the flask to remove the algal film on the sides of the flask, how would that have affected your results?

5. Estimate the total amount of your pollutant entering your community's sewage system each week. While you cannot determine this amount precisely, you can make several assumptions to estimate a range. For example, in your community, how much shampoo goes down the drain and into the sewage system each week? Measure how much shampoo you use each time you wash your hair, then multiply the amount by the number of times you wash your hair each week. If you assume that everyone else washes their hair as frequently as you do, then multiply the amount of shampoo you use per week by your community's population size. Show your calculations below.

6. Does using your community's entire population size make your estimate of shampoo quantity higher or lower than what the real value may be? Explain some assumptions you made that were necessary to do the calculations. Which assumptions would result in overestimated values, and which would result in underestimates of the true amount?

7. Discuss potential food chain effects. How could large quantities of your pollutant affect local aquatic species in your area, such as filter feeders, larval fish, and benthic scavengers?

8. Is it valid to extrapolate the results of this experiment to make predictions about what effects your chemical might have in the ocean? Discuss why or why not. If not, describe how you would design and perform another study that would provide better information to answer your question.

WEB EXPLORATIONS FOR FURTHER STUDY:

U.S. Geological Survey Toxic Substances Hydrology (Toxics) Program
http://toxics.usgs.gov/index.html
Initiated in 1982, this program conducts research on pollution and contamination affecting aquatic ecosystems throughout the country and provides cleanup and waste-disposal strategies to reduce future risks of contamination.

- **Activity**: Visit the homepage and, under the *Research Projects* column, click on the *Watershed and Regional Scale Contamination* link. On the next page, click on the *Emerging Contaminants in the Environment* link to see information about PPCPs in aquatic environments. Prepare a report describing the current knowledge concerning endocrine disruptors in aquatic environments. What are endocrine disruptors, and what ecological problems can they cause? What types of animals are most affected by these chemicals? Where have scientists detected endocrine disruptors? What types of PPCPs have been detected in water bodies in your state?

The Environmental Protection Agency
http://www.epa.gov/ppcp/
This site is a clearinghouse for PPCP environmental research.

- **Activity**: An excellent illustration showing the origins and eventual fate of PPCPs in the environment is found at **http://www.epa.gov/ppcp/pdf/drawing.pdf**. Trace the path of an antibiotic excreted by a farm animal to the habitat where the chemical ultimately ends up.

Northwest Fisheries Science Center, Harmful Algal Blooms Program
http://www.nwfsc.noaa.gov/hab/index.html
This branch of NOAA has been conducting research on harmful algal blooms (HABS) in the Pacific region for over 50 years.

- **Activity**: Visit **www.nwfsc.noaa.gov/hab/outreach/education.html** and watch the video "Science at Sea and the Hunt for Killer Algae". Prepare a report that ties together the content of the video with concepts that you have learned in this laboratory exercise.

Additional Web research topics:

1. What is the difference between primary and secondary wastewater treatment? Does your community utilize primary or secondary sewage treatment? Go online to your community's wastewater management website to find out.

2. What are ways to reduce your usage of personal care products and/or what are alternative sources that are more environmentally friendly? Go online to find 3-4 alternative sources to common household cleaning products and/or personal health/beauty products.

Name: ____________________
Date: ____________________
Section: ____________________

WEB EXPLORATION REPORT

Topic: ____

LABORATORY 9

Phycology: The Study of Marine Algae

INTRODUCTION:

Have you ever been to the seashore and noticed algae piled up in the **wrack line**? For most people, this is their only direct exposure to a group of interesting and ecologically important photosynthetic organisms that are vital sources of food and shelter for numerous other marine species. **Macroscopic** algae (those large enough to be seen by the naked eye) are the seaweeds.

Many people refer to algae as marine plants, but although many types of seaweeds resemble plants, they are actually classified within the Kingdom Protista, and not in the Kingdom Plantae. Seaweeds are not considered plants because of key biochemical and physiological differences and they lack the complex tissues found in vascular plants to conduct water and food throughout the organism. As in land plants, algae have chlorophyll *a* and various **accessory pigments** to trap additional wavelengths of light, increasing their photosynthetic ability underwater. These pigments form the basis of traditionally classifying marine algae into 3 main groups, or **Divisions**, corresponding to their color (whether the seaweed is mainly green, red, or brown).

Seaweeds reproduce asexually by fragmentation or by production of spores that develop into offspring that are genetically identical to the parent. Additionally, gametes (eggs and sperm) are produced and unite in the water to form a new, genetically unique individual.

The goals of this laboratory are for you to be able to:

- Identify several types of common seaweeds from the intertidal zone and nearshore subtidal environments.
- Recognize the structural diversity within the Divisions.
- Explain the value of several adaptations of various algae.

I. OVERVIEW OF MAJOR GROUPS

1. Division Phaeophyta (Brown Algae)

Brown algae are the largest types of seaweeds and are seen mainly on rocky shorelines in cold waters. These seaweeds have numerous shapes and forms; some—like the Pacific coast giant kelp *Macrocystis*—grow to over 40 meters in length, forming dense underwater canopies, while others—like the rockweeds (*Fucus* and *Pelvetia*)—grow less than 0.5 m in height on hard **substrates** in intertidal regions. Many types of brown algae possess flotation structures called **pneumatocysts**, gas-filled bladders that keep the organism extended and growing upward toward the surface. Bladders like these keep the planktonic brown alga *Sargassum* afloat in the Sargasso Sea, located in the central North Atlantic Ocean. There, huge drifting mats of the seaweed create a unique habitat supporting a community full of cryptically-camouflaged animals. Periodically, this alga washes ashore on the lower East Coast (Figure 1).

Brown algae possess 2 forms of chlorophyll: *a* and *c*. Species vary in coloration from olive-green to dark brown because the green chlorophyll pigments are usually covered by several other accessory pigments, namely, carotenoids, xanthophylls (ZANTH-thu-fills), and fucoxanthin. These pigments capture light for photosynthesis even in dimly-lit coastal waters at 30 m depth.

Some species of kelp (brown algae in the family Laminariales) are eaten directly by humans. These include two staples of Japanese cuisine: wakame (found in miso soup) and kombu (used as a vegetable and as seasoning). But the greater economic value of kelp is in **algin** (AL'-jin), a substance commercially extracted from the cell walls and used to thicken and maintain the smooth consistency of foods, cosmetics, pharmaceuticals, and industrial compounds.

a. Examine the specimens of brown algae available to you and compare their structure to Figure 2. Identify the rootlike **holdfast**, consisting of numerous **haptera** that attach the alga to the substrate. The stemlike structure is the **stipe**, to which the flattened **blades** attach. The entire body of the alga is called the **thallus** (plural **thalli**).
b. Gas-filled pneumatocysts may be present on your specimens. On the Pacific giant kelp *Macrosystis*, a pneumatocyst is attached at the base of each blade, while only one giant pneumatocyst is found on the Elkhorn kelp *Pelagophycus* (Figure 3).

Figure 1. The planktonic brown alga *Sargassum natans*. Arrow shows one of the pneumatocysts for flotation. The lacy skeletons of encrusting bryozoans (*Membranipora membranacea*) are visible on several blades and pneumatocysts. Miami Beach, Florida.

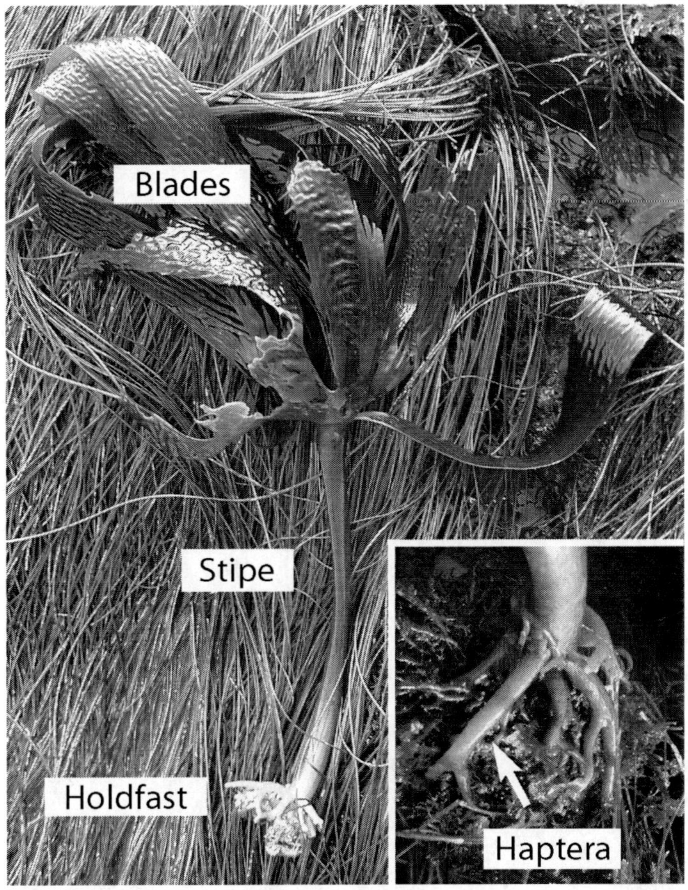

Figure 2. Major parts of a seaweed, shown on the southern sea palm *Eisenia arborea*. La Jolla, California. Size: 0.5 m.

Figure 3. Elkhorn kelp *Pelagophycus porra*, a brown alga commonly found washed ashore on Southern California beaches. This species forms canopies on the seaward edge of giant kelp (*Macrocystis*) forests, and the stipe may grow up to 30 m in a single year.

2. Division Chlorophyta (Green Algae)

Seaweeds in this group are usually bright green in color due to the presence of chlorophyll *a* and *b* (the same pigments found in land plants). Because of this, and also because green algae store their food in the form of starch (as do land plants), scientists believe that the first land plants evolved from a green algal ancestor more than 420 million years ago.

Green algae are more abundant in freshwater habitats of ponds and streams, but there are several important coastal marine species. Resembling sheets of green tissue paper, the sea lettuce *Ulva* is a **cosmopolitan genus** (one that is found throughout the world) in shallow intertidal regions (Figure 4a). It often grows abundantly in coastal areas with high levels of dissolved nutrients in the water *(i.e.,* polluted areas). Another widespread genus is *Codium* (Dead Man's Fingers), encountered on both temperate Pacific and Atlantic Coasts in both hemispheres.

On warmer subtropical and tropical shores, many **genera** of green algae deposit calcium carbonate ($CaCO_3$) throughout their tissues, giving them a crunchy texture. Examples of these genera are *Halimeda, Udotea, Penicillus,* and *Acetabularia.* When these **calcareous** algae die, their **thalli** break into tiny fragments and become part of the sediments. In some parts of the Florida Keys, most of the sand is made up of the calcareous remains of these green algae.

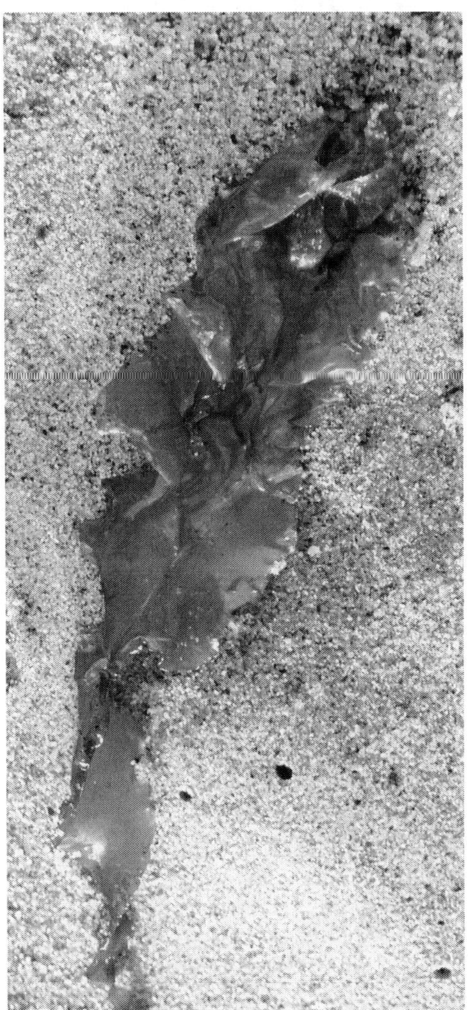

© P. Detwiler 2008

Figure 4. Two species of intertidal green algae found worldwide: a) Sea lettuce *Ulva lactuca,* Montevideo, Uruguay, left; size = 10 cm. b) *Codium fragile,* size = 20 cm, San Diego, California, above.

a. Examine specimens of green algae, noting the variety of thallus shapes.
b. If a fresh sample of *Ulva* is available, make a wet mount of a section of the thallus and observe it under the compound microscope. Do you see any tissue differentiation throughout the thallus? How many cell layers thick is this species? (Make a cross-section cut through the blade, if necessary.) Draw your observations below.

3. Division Rhodophyta (Red Algae)

This is the largest and most diverse group of seaweeds, in terms of the number of different species in this Division. Thallus morphology can vary from tiny tufts of filaments to large, shaggy sheets (Figure 5). Most red algae range in color from light pink to deep red, although many species are brownish-green and others, almost black. Despite these differences in coloration, all red algae contain chlorophyll *a* and *d*, and the accessory pigments **phycoerythrin** and **phycocyanin**. These pigments are called **phycobilins** and are reddish and bluish in color, respectively. These assist in the process of photosynthesis by absorbing extra wavelengths of light that chlorophyll *a* cannot, and thus, red algae are able to survive in deeper water than other types of seaweeds. However, red algae are found most abundantly in the temperate intertidal zone, forming a fuzzy "turf" on wave-beaten rocks.

Like green algae, some red algae are calcareous or **coralline**, and the thallus may appear either bushy or as a flat, **crustose** form growing over surfaces (Figure 6). In tropical reefs, crustose coralline algae play an important role in cementing the reef and stabilizing coral rubble.

Red algae in the genus *Gelidium* provides the useful substance **agar**, used to solidify liquids in foods and also to make substrates on which to grow bacteria in the laboratory. Another thickening substance commonly used in food products is **carrageenan**, obtained from the Atlantic species *Chondrus crispus.* Several types of red algae are eaten directly. The thallus of the temperate Pacific genus *Porphyra* (called **nori** in Japan) is dried and processed into a thin sheet for wrapping sushi, and *Palmaria* (known as **dulse** on the East Coast and in Great Britain) is eaten as a "sea vegetable."

Figure 5. The Pacific red alga *Chondracanthus exasperatus* ranges from Alaska to Baja California. Inset shows numerous papillae (bumps) on the surface of the blade. Its common name is "Turkish towel." Laguna Beach, California.

a. Examine any specimens of red algae. If coralline forms are available, inspect the tips of the branches for small, doughnut-shaped **conceptacles** with tiny pores in the center (Figure 6a, inset). Conceptacles house the reproductive structures that release gametes or spores through the pores and out into the water column.

b. Notice how the intertidal coralline algae are segmented with flexible joints along the stipe. What is the survival benefit of this adaptation? Consider the habitat of these algae in your answer.

Figure 6. Calcareous red algae: a) Bushy, articulated thallus of *Bossiella*. Arrow points to a reproductive conceptacle (inset); b) Delicate hair-like stipes of *Jania*; c) Crustose algae *Pseudolithophyllum* and *Lithothamnion* covering a rock. Arrows show thickness of the encrusting thalli. San Diego, California.

II. SEAWEED IDENTIFICATION

1. Obtain fresh or preserved seaweed specimens from your instructor.
2. Notice the color of each fresh specimen. Based on its color, to what Division does it belong?
3. On fresh seaweeds, note if there are any **epifauna** (either **mobile** or **sessile**) on your specimens. Place mobile organisms in containers of seawater for later observation under the dissecting microscope.
4. Examine your fresh seaweed for smaller species of **epiphytic** algae (algae that grow on other organisms). There may be crustose, filamentous, and fleshy species attached to your specimen.
5. Make a sketch of each seaweed in the following boxes. Use additional sheets of paper if necessary.
6. Use the identification keys provided by your instructor to identify the genus and species of your specimens.
7. When you have identified all the species using the key, check your answers with your instructor.

III. PREPARING A PRESSING OF A MARINE ALGA

If time permits, make an algal pressing using the herbarium apparatus.

> **DVD:**
> The techniques of pressing marine algae are demonstrated in the video clips for this laboratory.

Division:
Genus:
Species:

Division:
Genus:
Species:

Division:
Genus:
Species:

Division:
Genus:
Species:

Division:
Genus:
Species:

Division:
Genus:
Species:

LABORATORY 10

Invertebrate Zoology: Porifera

INTRODUCTION:

Among the simplest and most ancient multicellular animals are the sponges, members of the phylum Porifera (Latin for "pore-bearing"). Approximately 6,000 species known to science inhabit the oceans, with less than 200 species found in freshwater habitats. All sponges are sessile, and assume a variety of shapes: from massive, upright forms on the seafloor, to thin, encrusting growths over hard surfaces (Figure 1). Sponges diverged early from the evolutionary tree of animal life (Fig. 2), and no other animal phyla developed from them. Nevertheless, these simple animals play key roles in the ocean, from filtering massive amounts of seawater to housing diverse communities of **commensal** animals within the tunnels and canals of their bodies. Most people's familiarity with sponges is limited to synthetic products used for household cleaning, but in earlier times, natural sponges were commonly used as bandages, canteens, and even as toilet paper! Presently, sponges are being studied for new purposes: as potential sources of new drugs and chemicals to fight cancer and infectious diseases. So despite the lowly status of these ancient animals on the evolutionary ladder, they might soon provide humanity with dramatic medical breakthroughs, improving the lives of millions of people each year.

Figure 1. Dried specimens of Indo-Pacific sponges. Size: 10-15 cm.

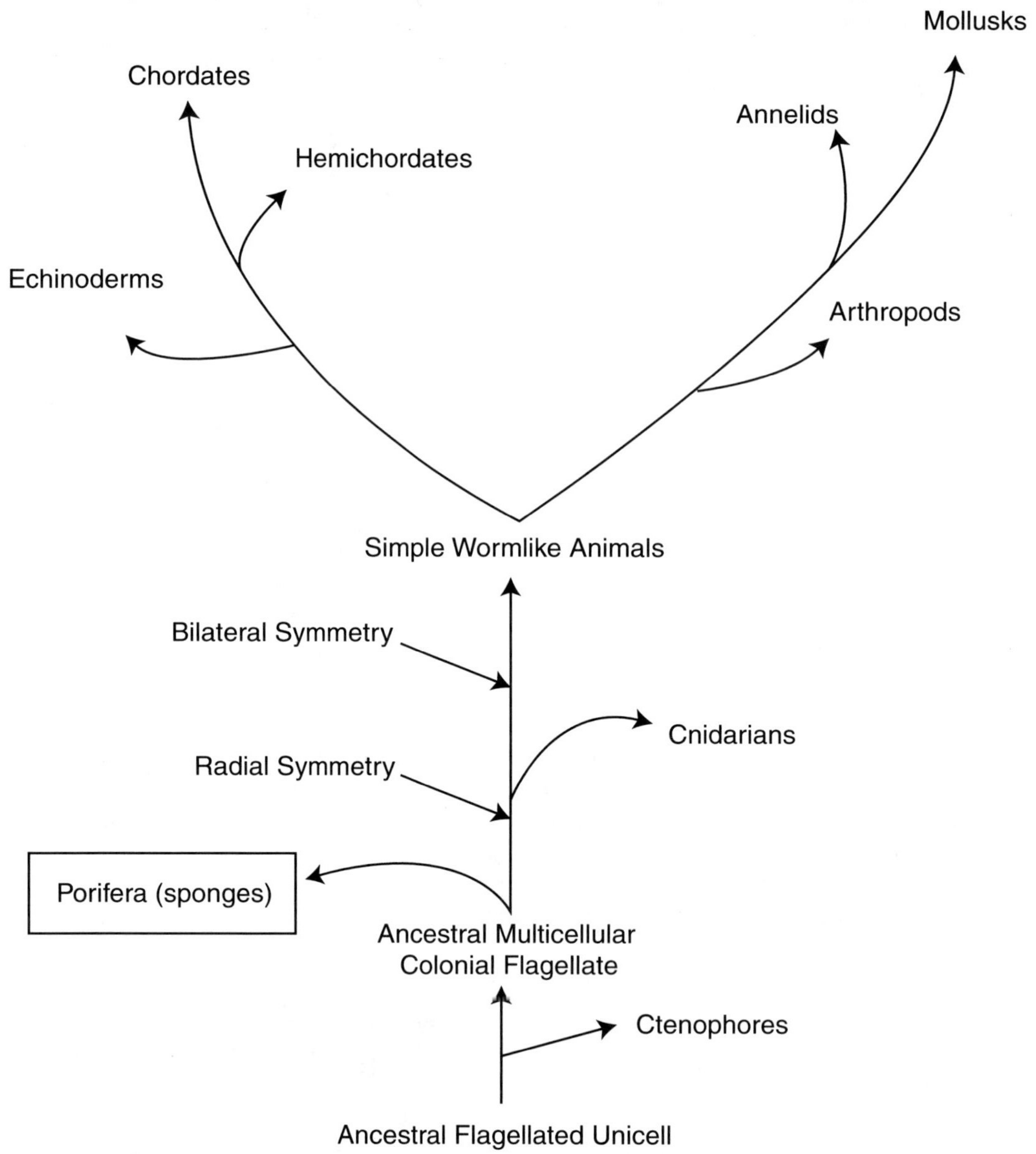

Figure 2. The animal "family tree": presumed evolutionary relationships among the major phyla in the animal kingdom.

The goals of this laboratory are for you to be able to accomplish the following:

- List the various cell types in the sponge body and describe their functions.
- Describe how sponges feed.
- Describe the support structures of the sponge body.

I. SPONGE FEEDING

1. Note the different kinds of sponges on display. Do you notice any kind of body symmetry?

   Can you identify a mouth?
   If there are any living specimens, do they respond to stimuli? Can they move?

2. How do sponges obtain food if they have no mouth and no ability to move? If living specimens are available, your instructor will attempt to demonstrate the unique feeding method of sponges by injecting dye into a sponge's body at various locations. Where does the dye come out? Draw a picture below, indicating the direction of water flow with arrows.

Water flow is produced by the action of **collar cells**, also called **choanocytes** *(ko-AN-o-sights)* lining the inside cavities of sponges (Fig. 3). On each cell, there is a large **flagellum** beating back and forth, and this movement pushes water into the **spongeocoel** (*sponge-o-seel*) (the central cavity of the sponge body). The spongeocoel opens to the outside via an excurrent hole, or **osculum** (plural ***oscula***) Replacement water is drawn into the sponge through incurrent holes called **ostia** (singular **ostium**) then through specialized tubular cells (called **porocytes**) that perforate the sponge's exterior surface. Small plankton and bacteria in the inflowing seawater are trapped within the collars of the choanocytes and ingested by the collar cell. Thus, water flow created by the animal brings food and oxygen to the sponge cells and removes wastes. What is this method of feeding called?

3. Prepare a **live mount** of a piece of sponge. Remove from a living specimen a small piece of material lining the interior of an incurrent canal and examine it under the compound microscope at medium power. Adjust the **iris diaphragm** (review parts of the microscope, Lab 1) to create contrast. Can you see the movement of flagella?

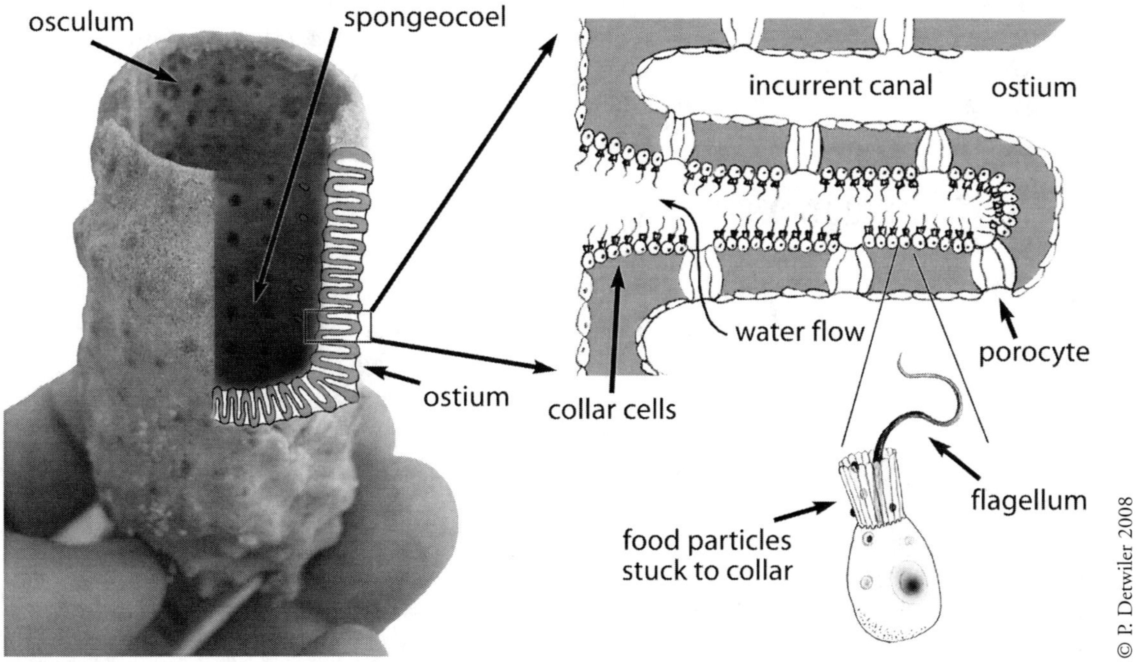

Figure 3. Cross section of a sponge showing direction of water flow through internal chambers.

Based on the previous information and your observations, what is the food of sponges?

In more advanced animals, food is broken down in the process of digestion in a specialized area of the body. Do you see digestive organs or a central stomach in the sponge?

So where does digestion of food particles take place in the sponge?

4. Filtration Rate. You can make a rough estimate of filtration rate by recording the length of time it takes for a known volume of dye solution to be completely expelled from an osculum. Multiply the volume of dye by the total number of oscula on the sponge to determine total volume of solution expelled per unit time. Convert to liters of seawater filtered per hour by using the following equation:

$$(A/B) \times C = \_\_\_\_\_\_\_\_ \text{ ml/sec} \times (3600 \text{ sec/1hr}) \times (1000 \text{ ml/1}) = \_\_\_\_\_\_\_\_\_\_ \text{l/hr}$$

where A = volume of dye injected (in ml)
B = the time it takes for all the dye to be expelled from an osculum (in seconds)
C = the total number of oscula on the sponge

DVD: If no living sponges are available, compute filtration rate of the sponge *Tetilla* shown in the video clip for this laboratory.

II. SPONGE REPRODUCTION

Sponges are **hermaphroditic** and so are capable of producing both eggs and sperm. In sexual reproduction, sperm cells are drawn inside the sponge body on the filtering current and fertilize the eggs held within. After a short development period, **meroplanktonic** larvae are released into the water column and later settle on a hard substrate to grow into the **benthic** adult form. Asexual reproduction can occur as budding or after fragmentation. Sponges are known for their regenerative abilities and are able to regrow after damage from predators.

III. SPONGE SKELETON

A sponge is not composed of just a mass of cells; cells produce certain structural materials to give support to the sponge body. These materials are either **spongin** (a flexible protein fiber), or spongin combined with tiny pieces of glass or calcium carbonate. Together, these materials compose what can be considered the sponge's "skeleton". To observe these skeletal elements (called **spicules**), you must first remove all the organic material from a piece of sponge.

Sponges are classified in part by the shape (**morphology**) of their spicules. Several different types are illustrated in Figure 4.

Some deepsea sponges create elaborate tubular skeletons of glass fibers (Figure 5a), while others make giant glass spicules to anchor themselves in soft sediments (Fig. 5b).

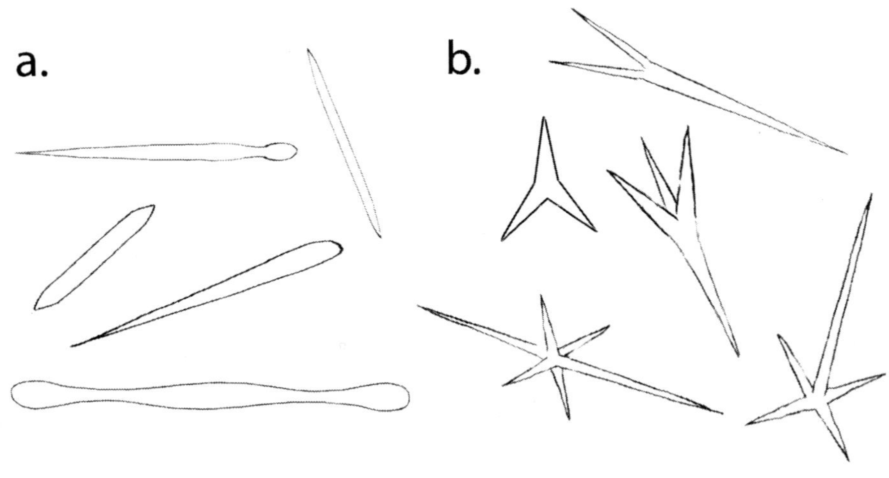

Figure 4. Examples of various types of spicules: a) monaxon (one axis); b) triaxons (three rays),

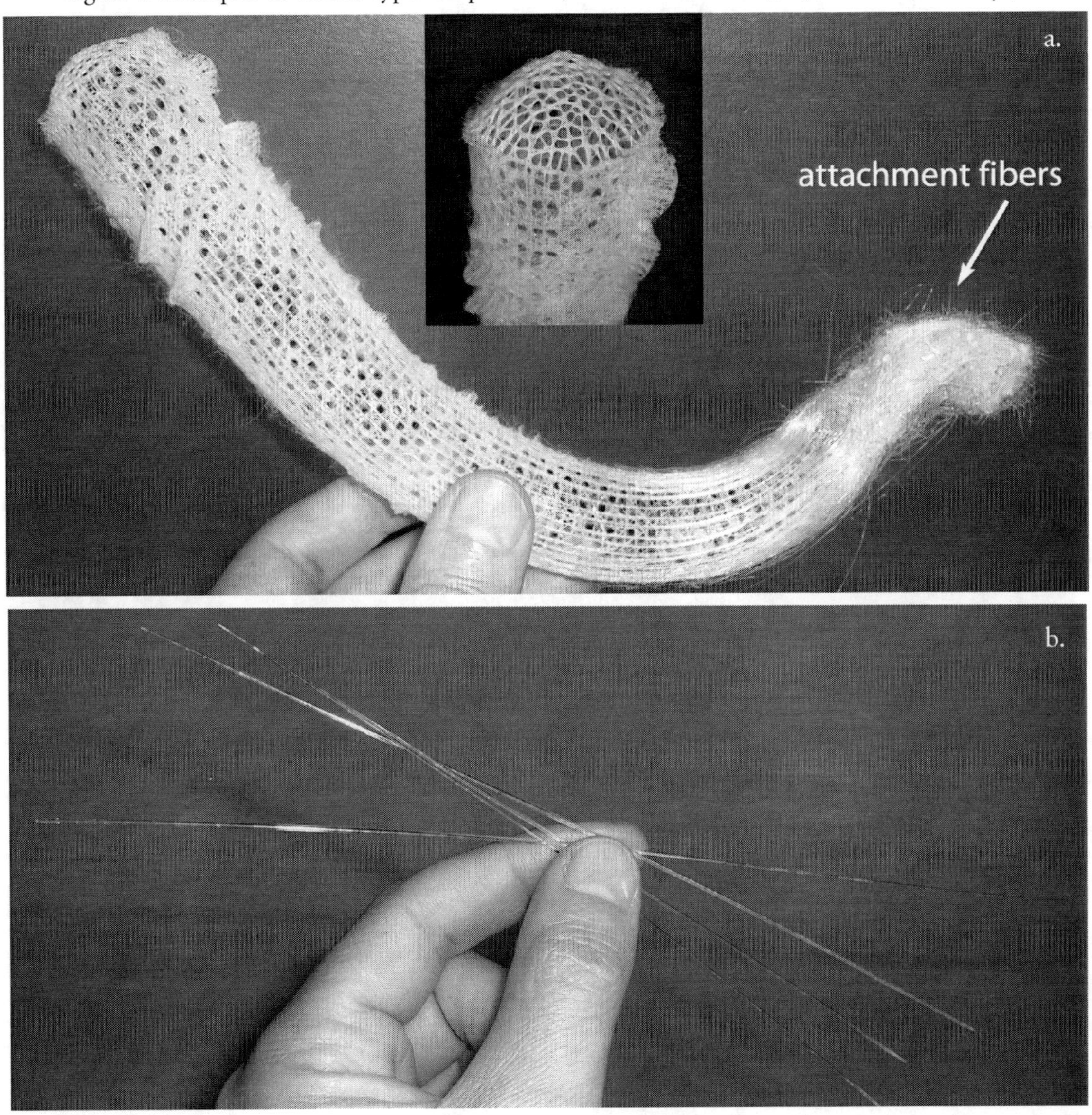

and tetraxons (four rays). Size = <0.05 mm.

Figure 5. Skeletal parts of deepwater glass sponges: a) Venus' flower basket sponge (*Euplectella*). The cleaned specimen shown above is prepared by dissolving away the cellular parts of the sponge with bleach, leaving the interconnected glass spicules. The animal attaches to the substrate with its long fibers. b) Spicules from the glass sponge *Staurocalyptus solidus,* collected by the author from 1800 m depth. Santa Barbara Basin, California.

Examine a prepared slide of sponge spicules, and draw them in the boxes below. If living sponges are available, isolate their spicules as follows:

1. Cut off a small piece (5 mm x 5 mm) of sponge and chop it into tiny bits with a razor blade. Use a glass slide as a cutting board.
2. Place the material into a test tube and add 5 ml of household bleach. Using tongs, hold the test tube over a flame and slowly heat to a gentle boil for several minutes. At all times, keep the test tube pointed away from yourself and your laboratory partners. Do not allow the test tube to boil over and spatter.
3. Place the test tube in a rack to cool. Then add enough distilled water to fill the tube halfway. Allow the residual material to resettle. Then pour off most of the water, keeping the residual material in the bottom of the tube.
4. Fill the tube again halfway with distilled water.
5. Pour the contents of the test tube through a piece of filter paper. Use a toothpick or metal spatula to collect some of the residue on the filter paper, and place a small amount of it onto a glass slide.
6. Add a drop of water and a coverslip, and observe under the compound microscope. Make drawings of the various types of spicules below, and identify which types of spicules are present from the examples given in Figure 4.

Species: Spicule type: Magnification: _______X	Species: Spicule type: Magnification: _______X

What are the spicules made of? Add a drop of I N HCL to a drop of the spicule residue, and watch for 15 sec. Do bubbles appear, indicating the formation of CO_2 gas?

If bubbles form, the spicules are calcareous. If no bubbles form, what material are the spicules made of?

Examine any dried sponges set out for demonstration. Can you tell by touch which kinds have skeletons of glass or calcium?

Not all sponges are filter feeders! Astonishingly, several species of recently discovered deepsea sponges are carnivorous predators. These sponges eat small crustaceans that collide into the sponge body and become trapped in the bristly external spicules that hook the tiny prey and hold them like Velcro against the sponge's body wall. Within a few hours of capture, the sponge grows a cage of longer spicules around the prey and begins digesting the helpless victim.

Name: ____________________
Date: ____________________
Section: ____________________

IV. REVIEW

Label the diagram of a cross section through a sponge wall. Indicate the features listed below, and show the direction of water flow with arrows.

choanocytes, porocytes, spicules, ostium, osculum, incurrent canal, spongeocoel, excurrent canal, flagella

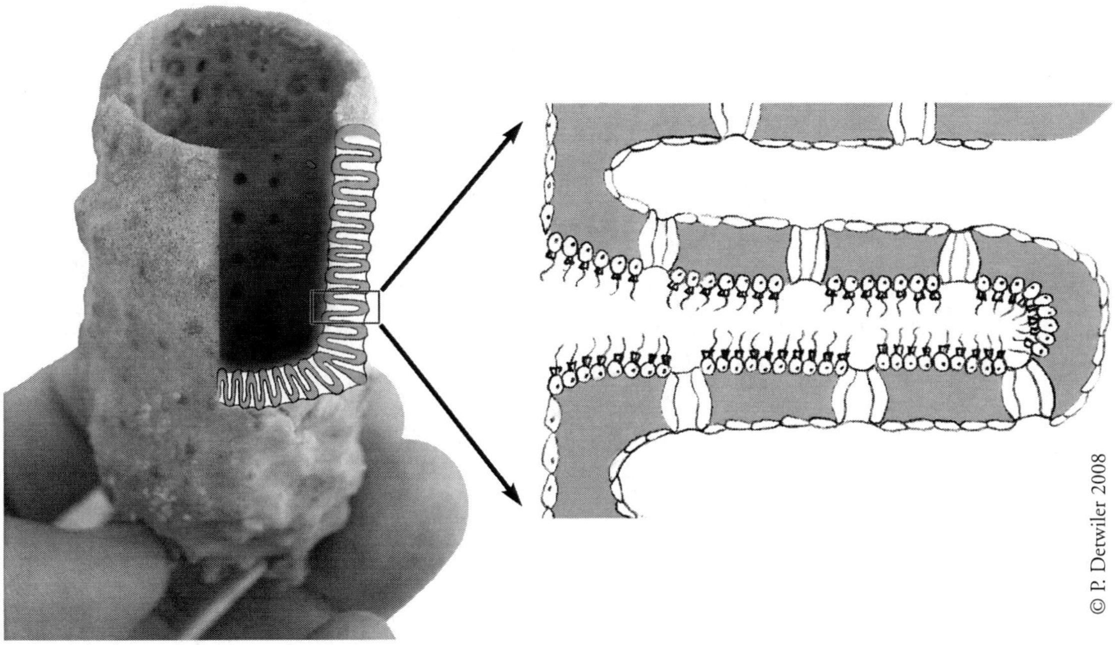

Figure 6. Cross section of sponge wall, magnified to show details.

Complete the paragraph by inserting appropriate terms encountered throughout this laboratory exercise.

The phylum Porifera (*porus* 'pore', *ferre* 'to bear') is made up of sponges—literally pore-bearing animals. The bodies of some of these unspecialized organisms may show ____________________ symmetry, but many encrusting forms lack symmetry. Additionally, sponges ____________________ organs, sensory systems, and muscle fibers. There are (*how many?*)____________________ major groups of sponges, generally corresponding to the material making up their skeletons. The individual skeletal elements (called ____________________) can either be composed of ____________________ or ____________________, or the sponge may lack both and instead have its body supported by a network of organic fibers made of the protein ____________________. Because there is no respiratory system, cells receive oxygen by the process of simple ____________________. The beating of ____________________ create a current to take in water through numerous ____________________ (tiny pores) into the sponge body. Food particles are captured on choanocytes (also called ____________________) and consist of phytoplankton, bacteria, and particles of organic matter. This method of feeding is called ____________________. Sponges generally live attached to hard surfaces and cannot move, so these animals are said to be ____________________. Sponges reproduce both sexually and asexually. A sponge can be torn into bits, and each bit can ____________________ into a new sponge.

V. QUESTIONS:

1. Do sponges have intracellular or extracellular digestion?

2. What roles do filter feeders play in aquatic ecosystems?

3. Describe the pathway of water through a sponge and how food is removed from the water. Use the terms *spongocoel*, *osculum*, *porocytes*, and *choanocytes* correctly in your description.

4. What are three ways that sponges are different from other sessile marine invertebrates?

5. How long would it take *Tetilla* (the sponge shown in the DVD video clip) to filter 120 liters of water, the average volume of a standard-sized bathtub? Show your work below.

6. What types of animals prey on spongs? Sponges are immobile, so how might they protect themselves from these predators?

WEB EXPLORATION FOR FURTHER STUDY:

Center for Marine Biotechnology and Biomedicine
http://sio.ucsd.edu/About/Research_Overview/Marine_Biotechnology_and_Biomedicine/
This research division housed at Scripps Institute of Oceanography is the world's largest coordinated biomedical research program devoted to exploring and developing new medicines and technologies based on marine life.

- **Activity**: Perform a Web search of the CMBB site or the Internet to investigate different types of drugs and chemicals derived from marine sponges. To start, use some, or any, combination of these search terms:

sponge compounds	sponge isolates	manoalide
antitumor	marine pharmacology	discodermolide
antiviral	marine drug discovery	crambescidins
marine medicine	secondary metabolites	

What species of sponges produce chemicals currently being investigated for use in treating diseases? Suggest possible reasons why these sponges produce such potent chemicals. How are these chemicals thought to work in fighting disease?

Name: ____________________
Date: ____________________
Section: ____________________

WEB EXPLORATION REPORT

Topic: ____

LABORATORY 11

Invertebrate Zoology: Phylum Cnidaria

INTRODUCTION:

The phylum Cnidaria consists of approximately 9,000 species that developed evolutionary innovations in their body structures allowing them to become efficient predators throughout the Ocean. These primitive animals are more complex than sponges and are characterized by several notable features:

1. A radially-symmetric body arrangement
2. A central mouth surrounded by a ring of tentacles
3. An incomplete gut (meaning the mouth is the only opening in the digestive system)
4. Specialized cellular microstructures used for capturing and stunning prey

One hypothesis for when these anatomical features may have first appeared is presented in Figure 1.

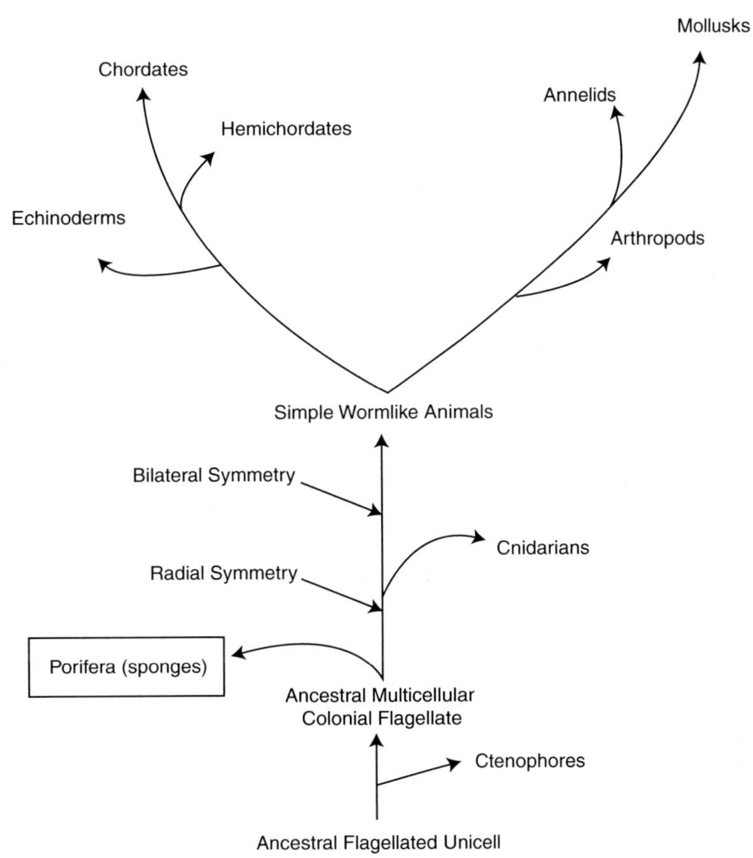

Figure 1. The animal "family tree": presumed evolutionary relationships among the major phyla in the animal kingdom.

A cnidarian may spend part of its life as a **polyp** or **medusa**. These two body forms are quite similar in appearance. The polyp exists as a benthic, sessile organism attached to a hard substrate, with its tentacles and mouth directed upward into the water column, while the medusa is planktonic, and the mouth and tentacles hang down from the body (called the bell). Both body forms consist of an outer layer of cellular tissue (**ectoderm**) and an inner layer of **endoderm**. Between these layers is a jelly-like material called **mesoglea** (mess-oh-*GLEE*-ah), which constitutes much of the weight of a jellyfish. Beachcombers may sometimes find blobs of transparent mesoglea washed up on the shore, the remains of jellyfish stripped of their living tissue layers by the abrasive sand and pounding waves in the surf zone (Figure 2).

The phylum contains 3 main classes:

Class **Hydrozoa**: The dominant form of the lifecycle is the polyp form, which produces planktonic medusae. Hydroids, fire coral, and **siphonophores** (sigh-*FAWN*-oh-fours) are members.

Class **Scyphozoa** (*SKY*-fo-*ZO*-ah): These are free-swimming jellyfish. The planktonic medusa is the dominant stage, while the benthic polyp stage is reduced or may even be absent.

Class **Anthozoa**: This group includes both soft and hard corals and sea anemones. The medusa stage is absent; these animals live as solitary sessile polyps or clonal colonies.

A fourth group, Class Cubozoa, consists of only 20 known species found in semitropical and tropical waters. The most famous of these is the Australian sea wasp, *Chironex fleckeri*—one of the most deadly sea animals in the ocean. Its venomous sting can kill a person within 3 minutes!

Figure 2. Mesoglea of the **pan-oceanic** moon jellyfish, *Aurelia aurita*, Miami Beach, Florida. You should not pick up any bits of jellyfish because stinging tentacles may still be attached. However, this specimen is completely stripped of the outer and inner tissue layers, and even if tentacles were attached, the sting of this species is generally too weak to be felt.

The goals of this lab are for you to accomplish the following:

- Describe the distinguishing characteristics of the phylum.
- Describe the adaptations for prey capture and feeding.
- Identify common species and recognize to which class each belongs.

I. HYDROZOANS (PHYLUM CNIDARIA, CLASS HYDROZOA)

1. Observation of *Hydra*.
   This cnidarian is found in freshwater, not the ocean, but provides an excellent example for studying the basic organization of the polyp body form.
   a. Use an eyedropper to transfer a small *Hydra* into a drop of water on a depression slide. (You don't have to cover it with a coverslip.) Observe its body and behaviors under the compound microscope.
   b. Do you notice any individuals that look like they have tiny hydras growing on them? These are **buds**, and this is one way that *Hydra* reproduces. The buds are genetically identical to the parent; therefore, they are clones. What method of reproduction is this called?
   Hydra can also form **gametes** (eggs and sperm cells) from within **gonads** in the body wall; these look like flat or swollen bumps. Examine the tentacles at 100X power. Notice the tiny bumps where the stinging cells (**cnidoblasts**) (*NI*-doh-blasts) are located.

   c. Gently touch one of the tentacles with a dissecting needle. Does the animal react?

   Therefore, do cnidarians possess nerve cells to respond to external stimuli?

   Would this imply that cnidarians also possess muscle fibers?

   These are evolutionary advancements of cnidarians and all other animals higher on the evolutionary tree. Indicate on Figure 1 where these features first appeared in the evolutionary history of the animal phyla. Do sponges have these features?

   d. To observe the feeding responses, put a tiny speck of red-stained yeast paste on the tip of a toothpick and add it to your *Hydra* specimen. Describe below how it responds to this food source. You may see a current flow circulating around the mouth, or the mouth may turn itself inside out, or the animal may sweep its tentacles back and forth over its mouth.

In cnidarians, food is ingested through the mouth, and digestion takes place in the central **gastrovascular cavity**. Wastes are also released through the mouth; there is no anus. Because of this, cnidarians are said to possess an **incomplete digestive tract**. At what point in the evolutionary history of animals did a complete digestive tract first develop? Indicate your hypothesis by drawing an arrow on Figure 1 showing where on the tree this feature could have appeared. Explain your reasoning below.

e. Draw a picture of *Hydra* below, labeling its **attachment disc**, **gastrovascular cavity**, **tentacles**, and any **buds** or **gonads**. Indicate where the mouth is located. If you have a prepared slide of a cross-section through the body of a hydra, draw it below and identify the **ectodermis**, **endodermis**, **mesoglea**, and **gastrovascular cavity**.

f. Transfer a *Hydra* into a drop of water on a flat slide, and add a tiny drop of methylene blue stain and a small drop of dilute (1-5%) acetic acid solution onto your slide. Cover with a coverslip, and scan the length of a tentacle at high power. What has happened to the stinging cells? Describe below:

The acid has caused the **cnidoblasts** (stinging cells) to discharge their contents, specialized microscopic structures called **nematocysts**, which are unique to the phylum Cnidaria. These tiny structures are used to capture food by harpooning and injecting prey with **neurotoxins** (poisons that attack the nervous system).

Find an undischarged cnidoblast, and look for a small, thorn-like spine on the outside edge of the cell. This is the trigger mechanism; when contacted, it causes the cnidoblast to fire, ejecting the barbed nematocyst thread into the unlucky prey (Figure 3).

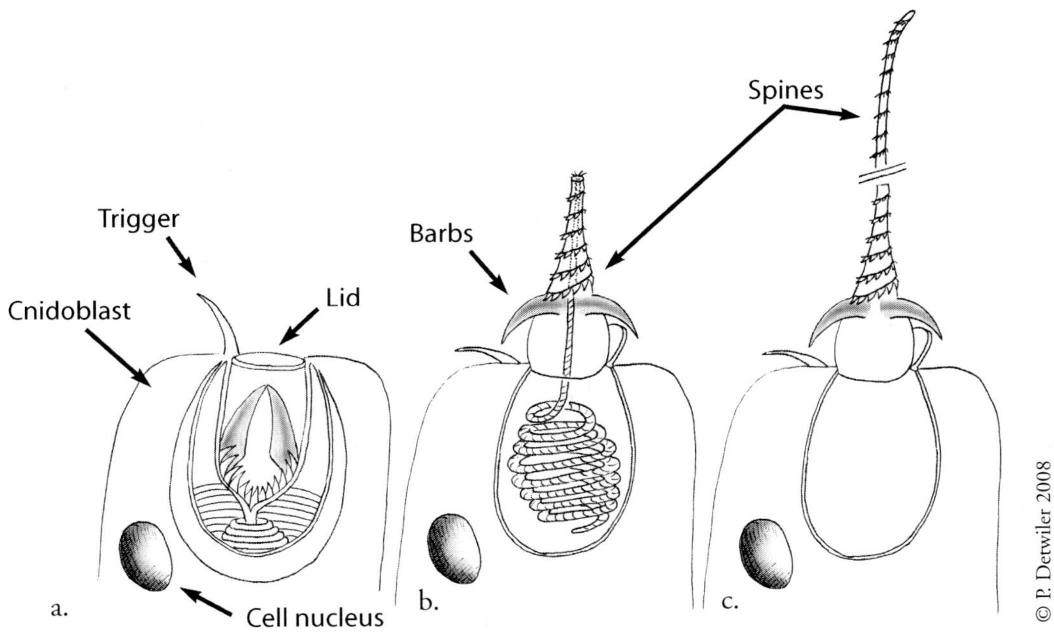

Figure 3. The structure and functioning of the cnidarian nematocyst. a) An undischarged nematocyst; b) Chemical or mechanical contact with the trigger instantly creates hypertonic conditions within the cnidoblast. Water immediately diffuses into the cell, and the increased water pressure squeezes the nematocyst, causing it to evert and pop out of the cnidoblast. c) A fully-discharged nematocyst. The coiled harpoon-like thread is often barbed to hold the prey.

2. Comparison of nematocyst structure and length.
   If other live marine cnidarians are available, prepare several slides of squashed tentacles to compare nematocyst structure among the different species.
   a. Use scissors and forceps to snip a bit of tissue off the tip of a tentacle and place it on a glass slide with a few drops of seawater. Cover with a coverslip and press down gently but firmly with your thumb to squash the tissue, causing the nematocysts to discharge. Examine under high power.
   b. Follow a discharged nematocyst along its full length. Estimate how many times longer the nematocyst is compared to the length of the cnidoblast that produced it. Note your observations in the table below:

Species	Nematocyst Appearance	Nematocyst Length (as a factor of cnidoblast length)

If brine shrimp larvae or small pond crustaceans (such as *Daphnia*) are available, add some to a fresh slide of *Hydra* and observe the feeding process.

II. COLONIAL HYDROZOANS

1. Observation of prepared slides of *Obelia.*
   Examine a prepared slide of *Obelia*, a colonial hydroid commonly found on pier pilings and other submerged surfaces. A colony consists of two types of polyps connected together by branching tubes. Feeding polyps are specialized for capturing food, so they have tentacles. Reproductive polyps lack tentacles, and are fed by nutrients delivered to them through the branching tubes. Identify both kinds of polyps on the slide.
2. Observation of preserved specimens.
   Other colonial hydrozoans known as **siphonophores** are planktonic and are best represented by the venomous **Portuguese Man-o-War** (*Physalia physalis*) and the harmless By-the-Wind Sailor (*Velella velella*, Figure 4). Each "animal" is actually a colony consisting of numerous polyps specialized for prey capture and defense, feeding, reproducing, and flotation. In *Physalia*, one giant polyp is modified as a gas-filled float (the **pneumatophore**) that keeps the colony buoyant at the surface, while *Velella* uses numerous gas-filled tubes beneath its papery sail for the same purpose. The Portuguese Man-o-War's tentacles may trail up to 40 m behind its float, and its potent nematocysts deliver an extremely painful toxin. In contrast, the stinging cells of *Velella* are incapable of penetrating the relatively tough skin of your fingertips, so these pretty blue specimens can be picked up and examined if you ever find one washed up on the beach.

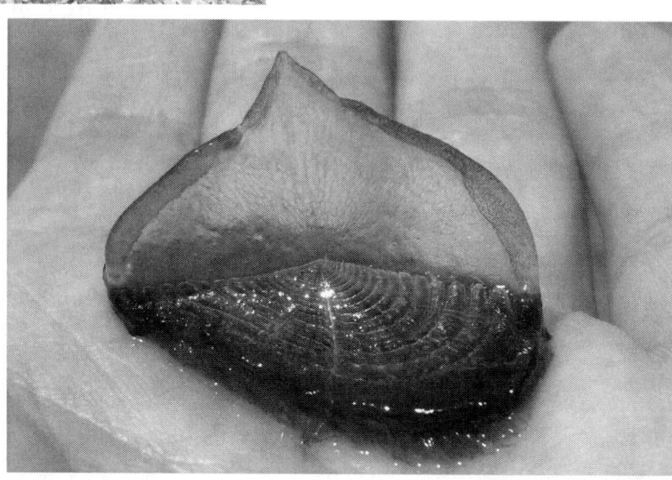

Figure 4. Pan-tropical siphonophores, washed up on the shoreline. Above: The pneumatophore of the Portuguese Man-o-War, Miami Beach, Florida; Right: By-the-Wind Sailor, Hermosa Beach, California.

III. CLASS ANTHOZOA

1. Observation of live sea anemones.
   a. If a live sea anemone is available, touch its tentacles gently; it will not hurt you. Describe what the tentacles feel like:

      The sensation is due to the nematocysts firing into your skin, but no pain is felt because they are not powerful enough to penetrate any deeper than your epidermis, the outermost layer of skin.

   b. Wait for the anemone to fully re-extend its tentacles, then pick up a small piece of raw seafood with forceps and touch it to the tentacles. Do not release the food from the forceps, but hold it against the tentacles for 1-2 minutes. How does the anemone respond to the food?

2. Observation of hard and soft coral specimens.
   Coral polyps are very similar in general form to sea anemones, but hard corals are colonies of interconnected polyps, with each polyp secreting its own calcareous skeleton to surround and protect itself. During the day, individual coral polyps are usually withdrawn into surface depressions called **corallites**, but at night, the polyps emerge, extending their tentacles to feed on plankton. Though individual coral polyps capture and ingest food, nutrition is shared throughout the colony via tissue channels connecting the corallites.
   a. Examine the hard coral skeletons provided by your instructor, and study the photos in Figure 5. Notice how the corallites on different species are differently shaped and have numerous partitions (septa) within them. Coral biologists use these features to identify species. Compare an anthozoan coral skeleton to a specimen of fire coral (a hydrozoan). What is an obvious difference?

   b. Soft corals (such as sea fans and sea whips) secrete a flexible skeleton of protein. Examine any dried specimens provided by your instructor.

Fire coral is not considered a true coral since it is not a member of the Class Anthozoa. Fire coral is actually classified as a type of calcareous Hydrozoan. The polyps of fire coral are extended during the day, giving the colony a fuzzy appearance. Anyone who scuba dives in tropical waters should recognize these colonies due to their intense, painful sting if touched by mistake. The author of this manual can tell you it is like being burned with a blowtorch, then having acid poured on the burn!

Anthozoans reproduce by both sexual and asexual reproduction. Eggs and sperm produced by gonads leave through the mouth during **spawning** (Figure 6) and fertilization occurs in the water column, resulting in a ciliated larva called a **planula** (Fig. 6c). The planula eventually settles to the bottom and metamorphoses into a benthic polyp. Coral polyps also asexually divide and grow outward, expanding the size of the colony. Any coral fragments broken off a parent colony may potentially become re-established in a new location and grow into a new colony.

Figure 5. Cleaned skeletons of Indo-Pacific and Caribbean corals, showing a variety of corallite forms: a) Rose coral, *Manicina areolata*, a solitary coral composed of only one polyp with noticeable septa. The mouth lies within the central groove; b) Golfball coral, *Siderastrea radians*, showing flattened corallites with numerous septa; c) Lace coral, *Pocillopora damicornis*; d) *Acropora serialis*. Asterisks indicate a large tubular corallite at the tip of each branch, a characteristic of this genus; e) The Caribbean fire coral, *Millepora alcicornis*.

a. b.

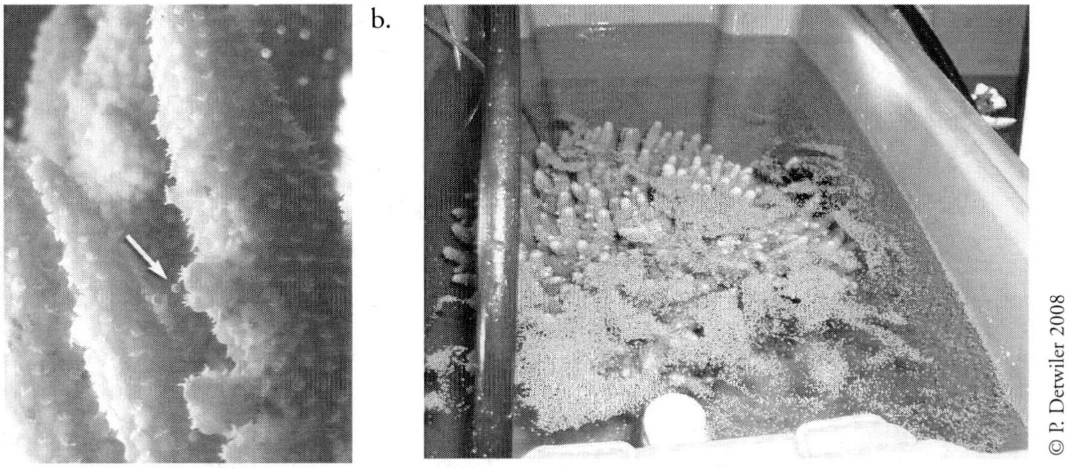

c.

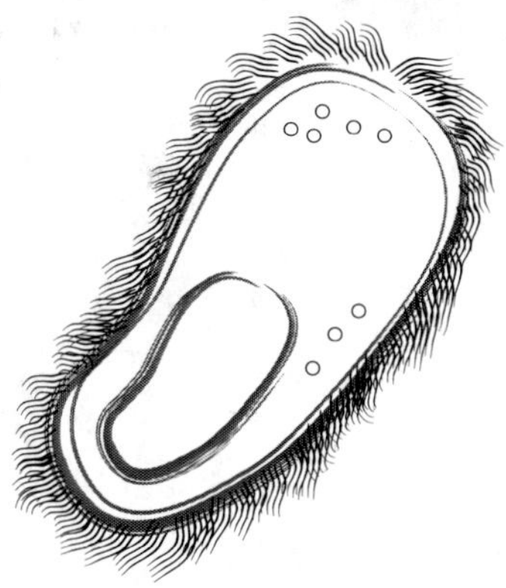

Figure 6. Sexual reproduction in the Indo-Pacific hard coral, *Acropora digitifera*: a) During spawning, coral polyps release bundles of eggs and sperm through the mouth (arrow). The buoyant bundles float to the surface where they rupture, releasing gametes to combine with those from another colony; b) All polyps in a colony can spawn within 2-3 minutes; this specimen has spawned in an aquarium tank. Lizard Island, Queensland, Australia. c) Planula larva of cnidarians. Size = 0.1mm. Illustration (c) M. Kwon.

Many types of hard corals are reef-building species, meaning their colonial skeletons grow massive enough to create a habitat supporting an entire community of species. There are 3 main types of reefs:

Fringing reefs are in shallow water, close to the shoreline, usually within swimming or wading distance.

Barrier reefs are further offshore, separated from the coast by a channel deep enough for large-boat traffic. The largest barrier reef is the Great Barrier Reef in the Coral Sea, along the northeast coast of Australia. Next is the barrier reef of Belize, then the Florida Keys.

An **atoll** is a ring-shaped coral reef encircling a central lagoon. Charles Darwin first proposed how atolls form. First, they begin as fringing reefs along the shoreline of oceanic islands. Over the course of millions of years, the islands slowly erode and sink beneath the sea surface, but the corals manage to keep pace by growing upward, staying within the shallows. Eventually, the island sinks completely, leaving the characteristic ring shape of the reef.

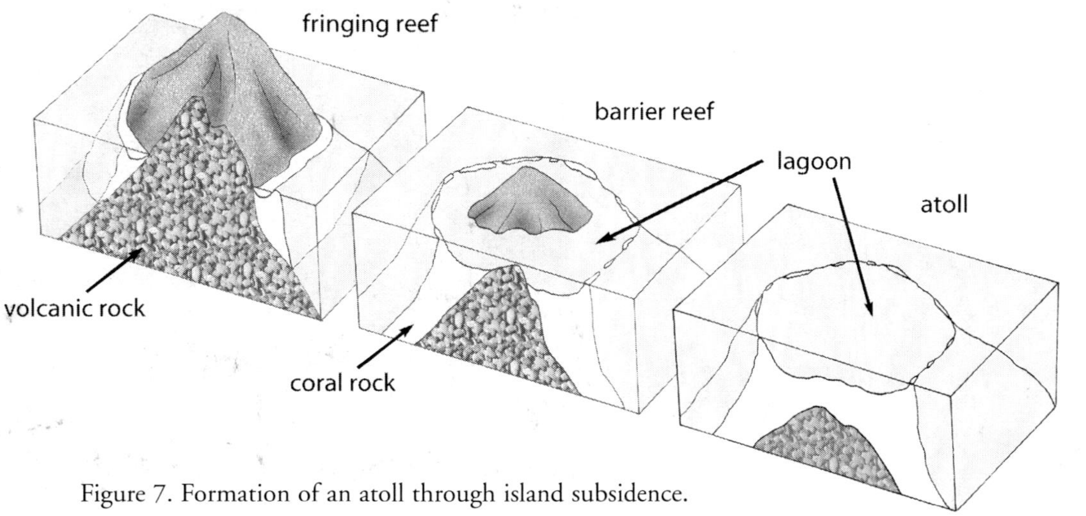

Figure 7. Formation of an atoll through island subsidence.

IV. CLASS SCYPHOZOA

This class contains the true jellyfish. Some species attain great sizes: the bell of the lion's mane jellyfish *Cyanea capillata* can grow up to 2.5 meters in diameter with tentacles up to 36 m long!

Examine a slide of a medusa or a preserved specimen of the common moon jellyfish, *Aurelia aurita.* This pan-oceanic species is found throughout the Ocean, sometimes forming large shoals along coastlines.

Notice the short, thin tentacles along the margin of the bell, and the structures that occur in 8 indentations spaced evenly around the margin of the bell. These indentations contain sense organs for detecting light and chemicals, and for helping the organism maintain its balance as it swims. The mouth is located in the center of the bell, at the end of a stubby stalk (Fig. 8). Around the central mouth are 4 oral lobes lined with nematocysts. Prey that touch these lobes are stung and swept up through the mouth into the gastrovascular cavity. Notice the numerous radial canals that extend in narrow lines from the center of the bell out to its margin. These radial canals connect to the gastrovascular cavity. What might be their function?

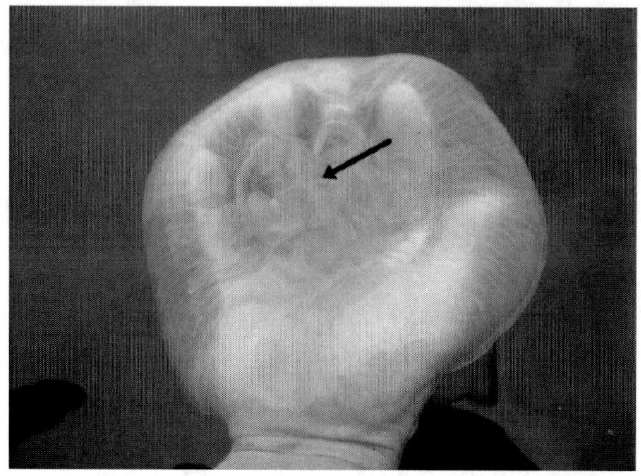

Figure 8. Underside of moon jellyfish, showing radial canals. Arrow shows the location of the mouth, surrounded by the oral lobes.

Sexes are separate, with four pink testis or ovaries surrounding the mouth (Fig. 9). During spawning, males release sperm cells through the mouth out into the water, and sperm enter the mouth of a female to fertilize the eggs internally. Zygotes develop into ciliated planula larvae, and escape from the female to enjoy a short-lived planktonic phase. The larvae then settle onto the seafloor and each develops into an asexually-reproducing clonal polyp called a **scyphistoma**. The scyphistoma produces tiny medusae (called ephyrae) by budding; each tiny ephyra later develops into the adult jellyfish.

Examine prepared slides of planulae, ephyrae, scyphistomae, and study any diagrams of the life cycle of *Aurelia.*

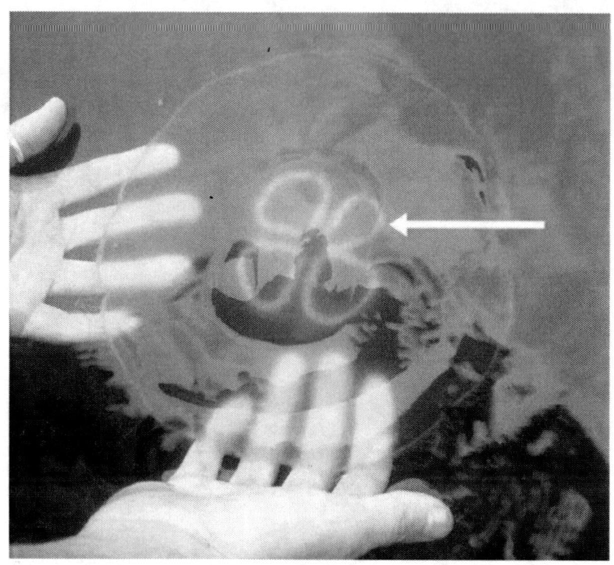

Figure 9. Top view of moon jellyfish floating just beneath the surface of Mission Bay, San Diego. Arrow shows the horseshoe-shaped gonads.

Name: ___________________
Date: ___________________
Section: ___________________

Topic: ____

WEB EXPLORATION REPORT:

LABORATORY 13

Invertebrate Zoology: Marine Arthropods

INTRODUCTION:

The phylum Arthropoda (*arthro* = joint, *pod* = foot) is the largest phylum of animals, found in every conceivable habitat in the ocean, as well as on land. Arthropods constitute nearly nine-tenths of all animal species recorded by scientists. The diversity of this group is staggering: over a million species have been described, with the majority being terrestrial insects. The defining characteristic of these bilaterally symmetric, segmented animals is their external skeleton (**exoskeleton**) composed of a durable carbohydrate and protein material called **chitin** (KAI-tin). To allow movement, the exoskeleton has joints of thinner chitin. This protective covering was the key adaptation contributing to the evolutionary success of this group, as the exoskeleton became modified into specialized appendages allowing walking, swimming, sensing, flying, burrowing, feeding, and respiring. Body growth occurs only when this covering is shed, or **molted**.

While insects are the dominant arthropods on land, only a few types of insects, the **neustonic** (new-*STAH*-nik) water striders of the genus *Halobates*, are found in the oceanic realm. Other minor groups of marine arthropods are the sea spiders (Class Pycnogonida) and the horseshoe crabs (Class Merostomata), described in the following sections.

The dominant marine arthropods are in the Class **Crustacea**, represented by about 40,000 species of lobsters, shrimp, crabs, barnacles, isopods, amphipods, and copepods. Many of these animals share a typical body arrangement of 2 main divisions, the **cephalothorax** and **abdomen**. Body segments typically bear jointed appendages modified for feeding, locomotion, respiration, or reproduction. From the Alaskan king crab with its 2-meter leg span to planktonic copepods less than one-quarter of a millimeter in size, crustaceans are vitally important members of the oceanic ecosystem and many crustacean species support major fisheries for human consumption throughout the world. Because of their ecological and economic importance, it is necessary that anyone studying marine science develop an understanding of the general biology and anatomical features of crustaceans.

The goals of this exercise are for you to be able to:

- Locate and identify external and internal structures of a representative marine crustacean, the Atlantic blue crab.
- Describe the functions of its various anatomical structures.
- Explain how the crab's adaptations help it survive in its environment.
- Compare and contrast anatomical adaptations between primitive and more modified decapods.

DVD: Please review the DVD chapter for this activity before the day of laboratory.

EXERCISE 1. OBSERVATION OF MARINE CHELICERATES.

These primitive marine arthropods possess unique appendages (**chelicerae**) modified for grasping food and feeding. These animals have no head region, and unlike all other groups of arthropods, lack antennae. Chelicerates include the horseshoe crab (*Limulus polyphemus*), sea spiders, and the arachnids (land spiders, scorpions, ticks, and mites). Study the external features of any preserved specimens available to you as you refer to Figures 1 and 2.

1. Horseshoe Crab

The horseshoe crab has six pairs of appendages in the anterior body segment (the shield-shaped **prosoma**). The first pair, the **chelicerae**, resembles tiny pinchers. Look at their position in relation to the location of the mouth (Figure 1). Based on their position, what is their function?

Limulus plows through the upper layers of sand in shallow waters using its walking legs. Compare the tips of these appendages modified for locomotion. Which appendages provide the force necessary to propel the animal forward along the sea floor?

Notice where the mouth is located: at the base of the walking legs. Notice that the base of each walking leg is spiny. Together these **gnathobases** (*gnath* = jaw) grind together when the animal moves its legs, effectively "chewing" its food.

DVD: How are these book gills used to circulate water?

The second body region (the spiny-edged **opisthosoma**) bears flattened plates called **book gills**, used for gas exchange.

Notice the compound eye on each side of the prosoma and the smaller pair of median eyes in the center. These simple structures detect both visible and ultraviolet light.

The long spine, called the **telson,** is used by the animal to right itself if it is ever flipped upside down by waves.

2. Sea Spiders

Sea spiders are benthic, ranging in distribution from the intertidal zone all the way down to abyssal depths. There are about 1,000 known species. Intertidal forms are tiny (most under 1 cm in size), while the leg span of some deepsea forms may reach up to 30 cm, slightly longer than a sheet of binder paper. Intertidal sea spiders can be found crawling among seaweeds and over bryozoans and hydroids, their prey.

A sea spider is mainly all legs: the body typically bears 7 pairs, and in males, one pair is modified into **ovigerous legs** to carry egg masses laid by the female until they hatch. The posterior **abdomen** is reduced to a tiny stump. The anterior cone-shaped **proboscis** projects outwards between a pair of **chelicerae** and is used for sucking prey (sea spiders are detritus feeders and carnivores, preying on small, sessile invertebrates).

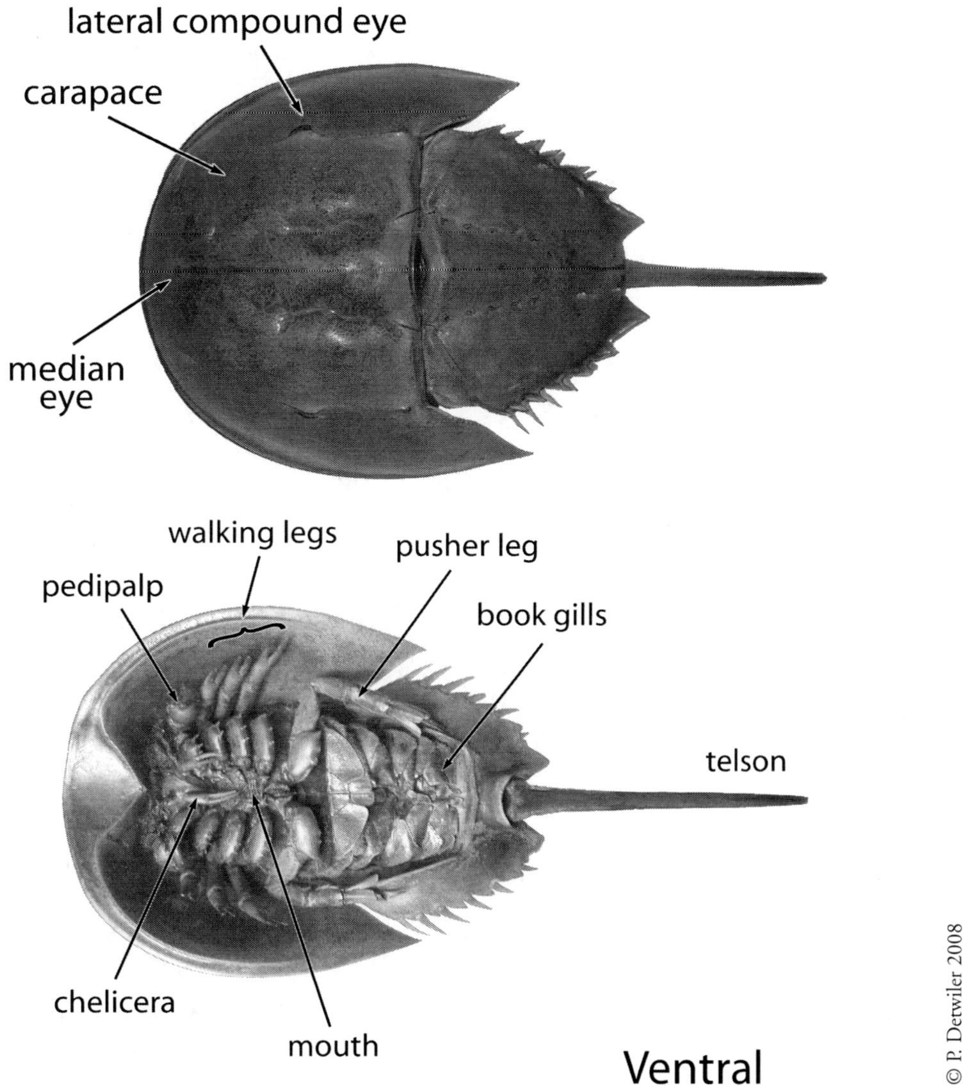

Figure 1. External anatomy of the Atlantic horseshoe crab, *Limulus polyphemus*: a) dorsal view, showing anterior prosoma and posterior opisthosoma; b) ventral view, showing appendages.

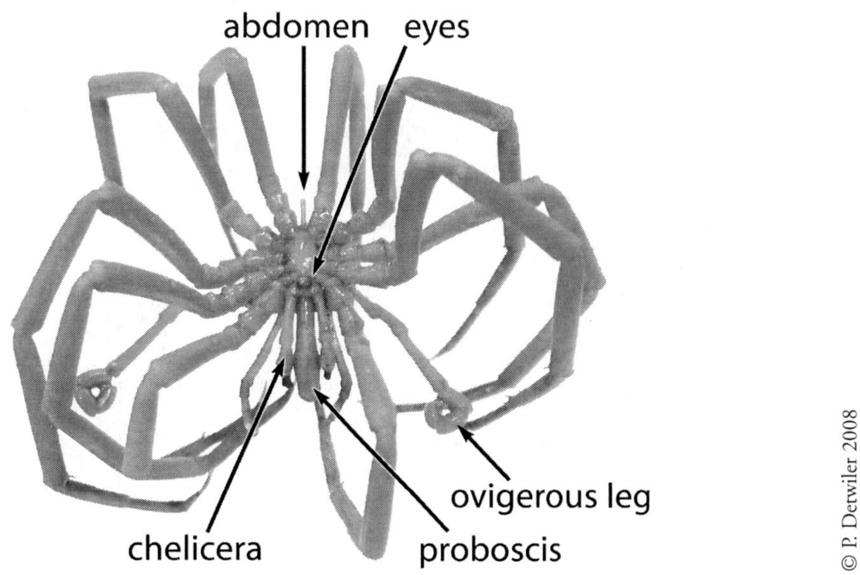

Figure 2. External features of a deepsea pycnogonid. Size: 18 cm.

EXERCISE 2: STUDY OF A REPRESENTATIVE MARINE DECAPOD.

Background:

The Atlantic blue crab (*Callinectes sapidus*) is a common member of the shallow water community on the East and Gulf Coasts. Female blue crabs release their hatched young in coastal regions near estuaries, and larval crabs drift as plankton upstream into fresher water, eventually undergoing metamorphosis into the benthic, adult form. Adults seasonally migrate downstream to the mouths of rivers to mate. Blue crabs have historically been the main source of income supporting generations of fishermen throughout the Chesapeake Bay region, though unfortunately, the adult population has been declining, and in 2006, the bay-wide harvest was the lowest since 1945.

You will perform a simple dissection of the blue crab. Before beginning this exercise, make sure that you understand the proper dissection and safety procedures as explained to you by your instructor.

External Anatomy.

(If you are examining a live animal, wrap rubber bands around the tips of its claws to prevent it from pinching. Your instructor will show you the proper way to safely handle a live crab.)

1. Body Regions

Examine the two major body regions: the **cephalothorax** and **abdomen**. The dorsal surface of the crab body is covered with a hard, flat **carapace,** with two long, lateral spines, while the abdomen is located on the ventral surface, tucked underneath the carapace (Figure 3).

2. Legs

Five pairs of walking legs (**pereopods**) are features of all decapod crabs; in fact, the order Decapoda is named for these 10 appendages. In crabs, the first pair of legs is modified into **chelipeds** (*KEE*-la-peds) not used for walking. In the Portunid family of crabs, to which the blue crab belongs, the fifth pair of pereopods is also modified.

How are the fifth pereopods different from the other three pairs of walking legs?

How do you think the animal uses these modified legs?

Are the size and shape of both chelipeds symmetrical to each other?

Are the teeth symmetrical on both chelipeds?

Which is the cutter cheliped? Which is the crusher cheliped? How can you tell the difference?

Is the cutter cheliped always found on the same side of all the crab specimens being examined today?

Make a sketch of both claws below, indicating which is which.

Is your crab missing any of its legs? Crustaceans can deliberately cut off their legs in a process called **autotomy** (*auto* = self, *tomo* = cut) when necessary to escape from predators. The next time the crab molts, the leg will begin to regenerate. After several molts, the leg will be fully regenerated. Are any legs in the process of regenerating?

3. Sensory Systems

Look around the crab's facial region. Sensory systems are usually clustered in the anterior region of the body, which is generally the first part of an animal's body to encounter the environment.

Compare the crab's eyes with the eyes of the Horseshoe crab. How are they the same? How do they differ?

Compound eyes are defining characteristics of arthropods. Remove an eye from a nonliving crab specimen and examine it under the dissecting microscope.

The compound eye consists of hundreds of individual light receptors called **ommatidia** (singular: **ommatidium**), each covered with a lens made of transparent cuticle material. Slice off the top layer of the compound eye, make a wet mount of it, and examine it under the compound microscope. Draw the arrangement of ommatidia below.

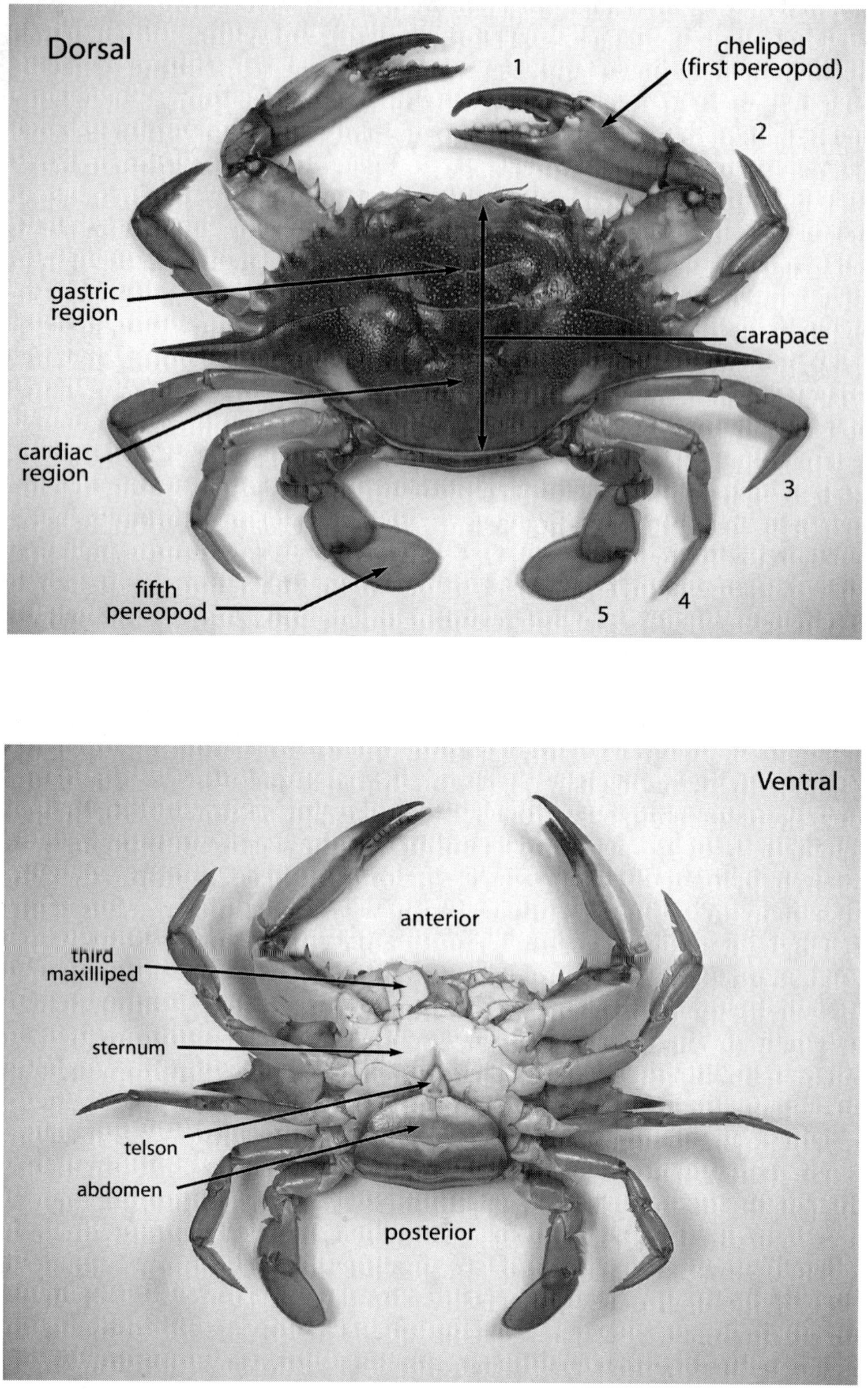

Figure 3. External anatomy of the blue crab.

What are some advantages of having compound eyes?

The antennae are sensory structures for detecting chemicals and for feeling (touch receptors). What happens when you gently touch an antenna on a living specimen?

How many pairs of antennae are present?

Inside the base of each short antenna is a tiny, gravity-detecting structure called a **statotocyst** that tells the crab which direction is up. Why is this important for a crab that swims in murky water?

In between the central pair of antennae is a pointed projection called the **rostrum**. On the blue crab it is tiny, but on other types of crabs the rostrum may be quite long.

4. Mouthparts

The mouthparts of decapods are complex. The first 3 pairs of mouthparts are the 3rd, 2nd, and 1st **maxillipeds**, respectively. These modified appendages collect and process food, then pass it to the hard and heavily calcified **mandibles,** where food is bitten into pieces. On a non-living crab specimen, examine the mobility of these appendages by moving each of them sideways with a probe to reveal the underlying mouthparts.

Find the smaller **second** and **first maxillipeds** located behind the 3rd maxilliped. The **second maxillae** are located behind the first maxillipeds. Behind the **first maxillae** are the mandibles. Move them with your probe. When the animal chews, how does the direction of movement of an arthropod mandible differ from that of a mammalian mandible (jaw bone)?

Examine the shape of the crab's mandibles. Are they adapted for chewing, grinding, or cutting food?

One of each of these paired appendages is shown in Figure 4.

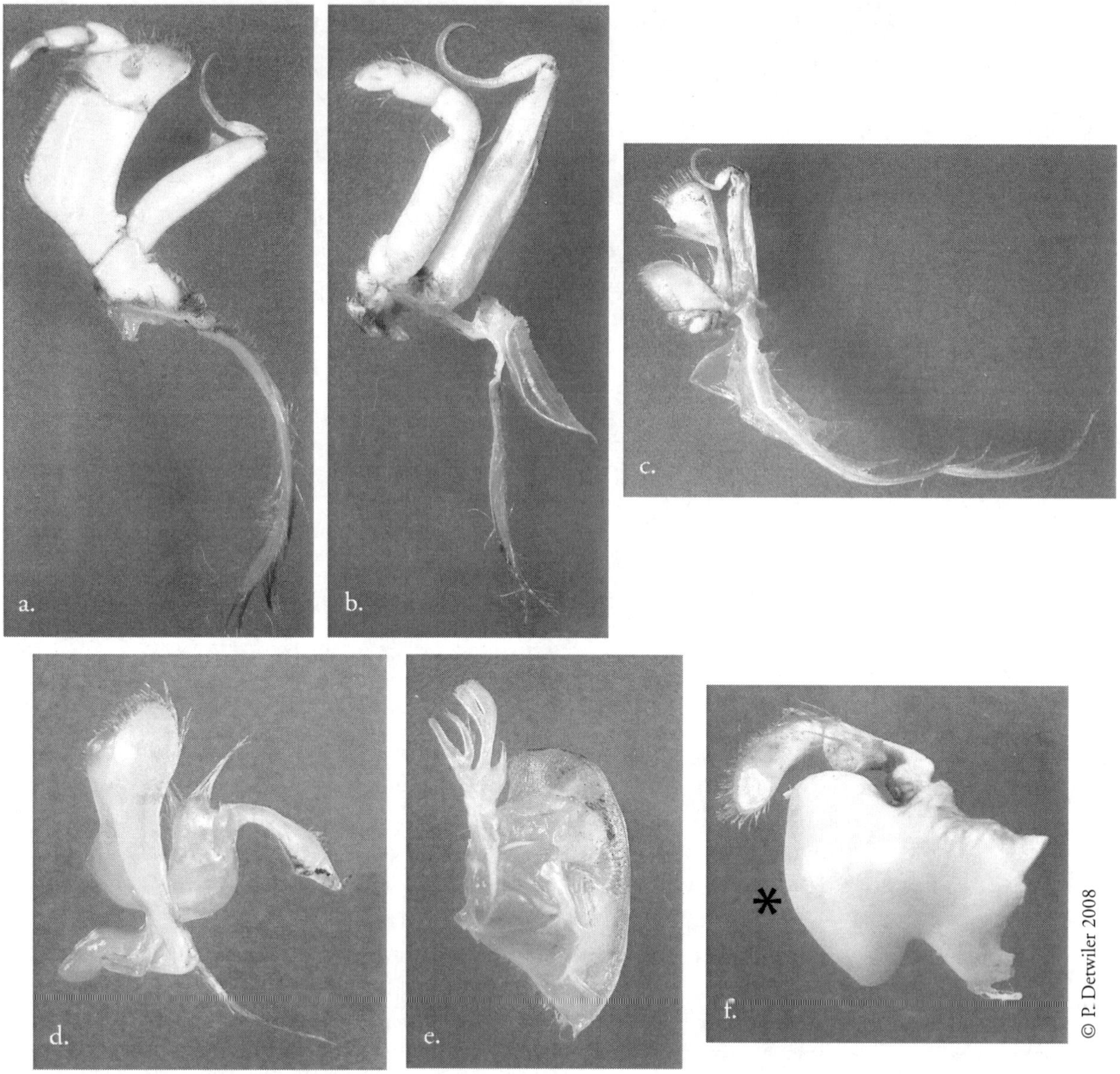

Figure 4. Mouthparts of the blue crab: a) 3rd maxilliped, bearing ventral gill cleaner; b) 2nd maxilliped, bearing a small gill and cleaner; c) 1st maxilliped, with dorsal gill cleaner; d) 1st maxilla; e) 2nd maxilla, with large, plate-like gill bailer that creates the water current to ventilate the gills in the gill chamber; f) mandible, bearing a palp on the dorsal surface (asterisk indicates the biting edge).

If you have a living specimen available in an aquarium, feed it a piece of mussel or fish and watch how the mouthparts are used to manipulate food toward the mandibles.

DVD: If no living specimen is available, watch the video clip of blue crab mouthparts in action.

5. Abdomen and External Reproductive System

The abdomen is curled underneath the cephalothorax and fits into a depression on the sternum (Figure 5). The abdomen is shaped differently in males and females, so this body part is **sexually dimorphic** (*di* = two, *morph* = shape or form). In females, the abdomen is wide and resembles the Capital Dome of Congress. In males, it is long and narrow, like the Washington Monument (Figure 6). What is the sex of your crab?

Uncurl the abdomen so it extends behind the animal. Look at the underside of the abdomen, and notice the tube running along its length, terminating at the **telson**, the last body segment. The tube may be transparent or dark-colored. What is this tube?

You will also see paired, feathery appendages attached to the abdominal plates. These are the **pleopods**, which are also dimorphic in males and females. Examine both male and female crabs to view the differences in these structures.

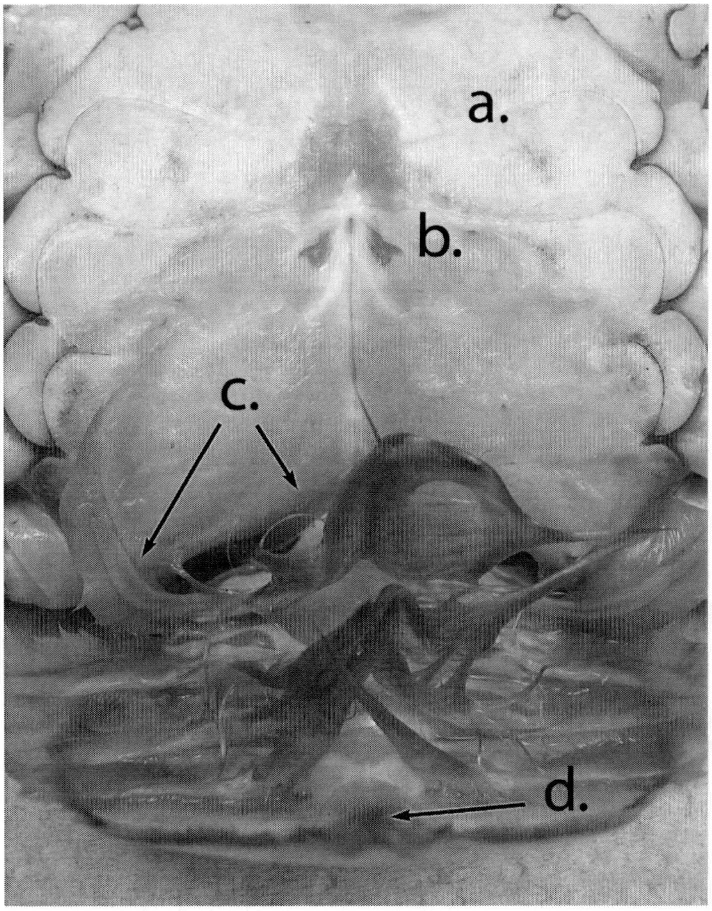

Figure 5. Abdomen of the female blue crab: a) sternum; b) gonopores; c) pleopods; d) telson.

Male crab: Two pairs of pleopods are modified into **gonopods**; the longer, curved first pair are sperm-transfer organs, inserted into the female during mating.

Female crab: Locate the **gonopores**, large, triangle-shaped openings on each side, under the abdominal plate at the level of the third pair of legs. Eggs are fertilized internally, then extruded from these openings and cemented onto the feathery pleopods. When a female crab is carrying eggs under her abdomen, she is said to be "in berry" (Figure 6). Larvae develop within the eggs and, at spawning, **zoea** larvae are released into the water column. These undergo further planktonic development into the **megalops** stage, which resembles the adult form, except that the abdomen is extended like a shrimp's (Figure 7). When the megalops finally sinks to the bottom, the abdomen curls underneath, and the adult form is achieved.

Figure 6. Preserved specimens of male and female blue crabs. a) male, showing narrow abdomen; b) female, showing eggs carried beneath the abdominal plate.

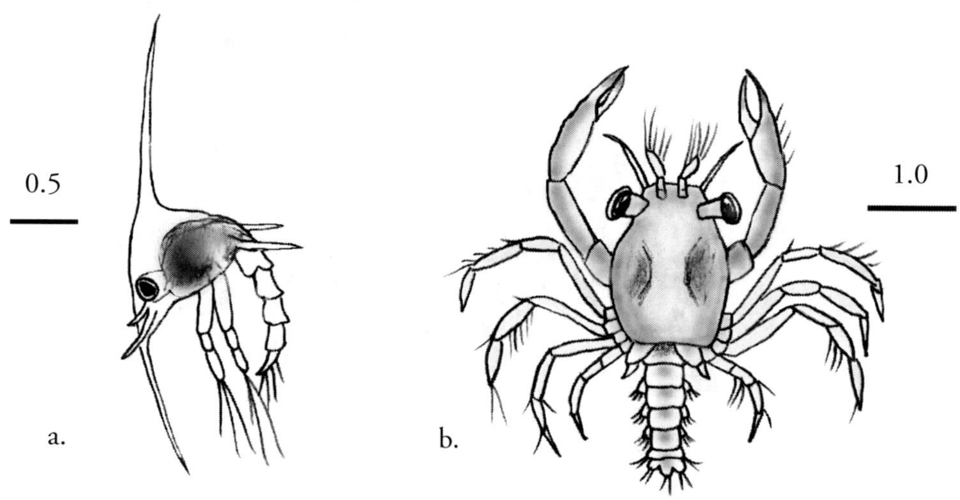

Figure 7. Larval stages of decapod crabs: a) zoea; b) megalops.

EXERCISE 3: INTERNAL ANATOMY

(NOTE: these descriptions are based on fresh, unpreserved specimens. If you are using preserved material, these descriptions of color and texture will differ from what you observe).

PROCEDURE:

1. Using kitchen shears or heavy scissors, cut the carapace along the inside of the dorsal margin, beginning at the rear of the animal and continuing forward.
2. Slowly lift off the carapace. It may stick to the underlying darkly-pigmented sheet of tissue (the **hypodermis**), so use your probe to gently separate it as you lift the carapace. Do not rip the carapace off roughly! The heart is connected to the dorsal body wall tissue, and it will tear if the carapace is pulled off.
3. Carefully remove all of the dark hypodermis with forceps to expose the internal organs.

Orient your crab with the anterior end facing away from you, and refer to Figures 8 and 9 to help identify the internal organs.

1. Overview of Internal Structures

The largest structure filling the anterior region of the crab's body cavity is the greenish/mustard-colored **digestive gland**, a highly branched extension of the gut. It extends throughout the body cavity, and fills the interior of both spines. It secretes digestive enzymes and absorbs nutrients. Lying on top of the gland are the **gonads**: in males, the **testes** resemble a white ribbon, while in females, the **ovaries** are orange-yellow and may completely obscure the digestive gland if the crab is reproductively mature. The stomach or **gastric mill** is centrally located in the anterior portion of the body cavity. Six bands of gastric muscles connect to it to the exoskeleton, and their contractions churn the stomach to grind the food within. The grayish **heart** is posterior to the stomach (Figure 8).

Blood enters the heart through openings called **ostia** in the heart's dorsal surface. Look closely; can you see them? When the heart beats, it propels the blood through vessels to the rest of the body, but the crab's blood vessels are open tubes that do not return to the heart. Blood simply sloshes through the tissues and eventually flows back to the heart. This is termed an **open circulatory system** and is a characteristic of arthropods. The oxygen-carrying molecule in the blood of crabs is the bluish pigment **hemocyanin**.

To see the rest of the internal structure, the digestive gland and the female gonad have to be removed. In fresh specimens, these tissues are soft and break into fragments easily, forming a mushy goop.

Using forceps, carefully remove these organs, piece by piece, without disturbing the gut and other underlying structures. Periodically rinse the body cavity gently with tap water to flush out soupy bits of gooey gonad and digestive gland tissues.

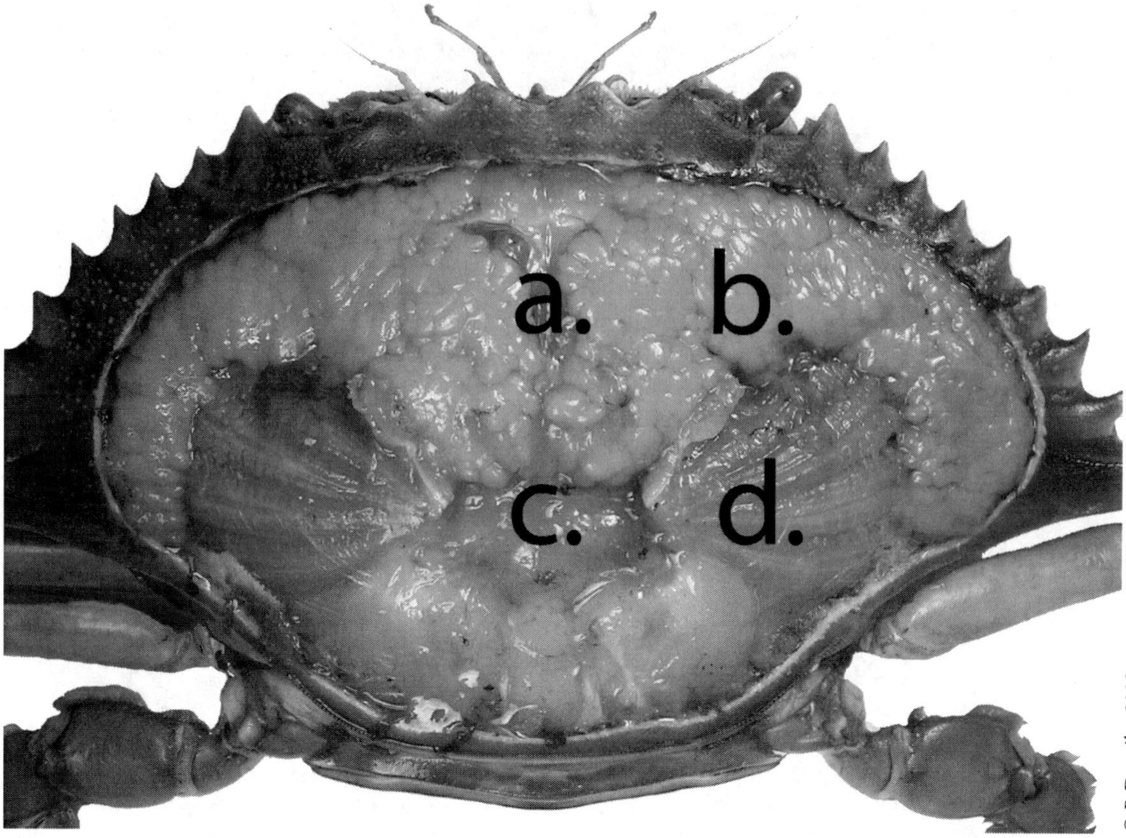

Figure 8. Superficial internal structures of the female blue crab: a) gastric mill (beneath ovary tissues); b) ovary; c) heart; d) gills.

2. Internal Reproductive Structures

If you have a male crab, trace the length of the testes to where they attach to the either side of the gastric mill. This coiled area is the **vas deferens,** where sperm formed in the testes is processed into **spermatophores**, tiny packages of sperm that are inserted into the female during mating. Remove a tiny piece of vas deferens tissue and view it under the compound microscope. Draw the spermatophores below:

If you have a female crab, you will see a pair of oval **seminal receptacles** on the floor of the body cavity. Before you removed the ovaries, they were joined to these structures. The male inserts spermatophores with his gonopods through the female's gonopores and into these receptacles, which store the sperm until the female lays her eggs.

3. Digestive System

As you recall, the 1st pereopod gathers or captures food that is passed by the maxillipeds and maxillae to the mandibles. Past the mouth, a short **esophagus** connects to the dorsal stomach. Note the gastric muscle bands connecting the stomach to the carapace. Imagine how the stomach would move in a living crab when these different bands contract.

1. Cut open the stomach with scissors and view the **gastric mill**, the food-grinding apparatus lining the inside surface. The gastric mill, along with most of the lining of the digestive tract, is made of chitin and is shed during each molt. Are there any stomach contents you recognize? Crabs are omnivorous scavengers, so there may be sediments and shell fragments. Rinse out the gastric mill, if necessary; then feel the inside surface. Describe the texture below:

2. Place a piece of the mill under the dissecting scope and examine the various teethlike structures modified for crushing, grinding, and straining. Draw a magnified view of the mill below:

3. If you have carefully removed the gonad, the intestine should have stayed connected to the stomach. Trace along its length to where it terminates, where undigested food exits the body.

4. Respiratory System

1. Examine the **gills** enclosed within the two **gill chambers** on both sides of the body cavity (Figure 9). Remove the thin membrane covering the gills, and you will see a thin, hairy, strap stretched over or along the gills. This is the **dorsal gill cleaner**, which sweeps debris and particles off the gill surfaces—an especially useful feature for crabs that inhabit muddy environments!

2. Remove all the gills from one side of the body to reveal the **ventral gill cleaner**.

3. The gill cleaners cannot move on their own; they contain no muscle tissue. So what structures are they attached to that control their sweeping movements? To discover this, grasp a gill cleaner with forceps and move it back and forth. What body part moves in response to this? Write your answers below:

   Dorsal gill cleaner attaches to:
   Ventral gill cleaner attaches to:

4. Place a gill in a dish of water and examine it under the dissecting microscope. Each gill is composed of hundreds of stacked sheets of thin tissue called **lamellae** that together provide a huge surface area for dissolved oxygen and carbon dioxide to diffuse across during gas exchange.

5. The crab is a large animal, and its gills are underneath a heavy exoskeleton. How can seawater circulate inside the exoskeleton to ventilate the gills for gas exchange? What structures could the animal use to generate a water current?

6. If you have a living crab in an aquarium, slowly squeeze a pipette of dilute food coloring/seawater solution in the vicinity of the crab's face, and watch for water currents.

> **DVD:** If no living specimen is available, water circulation is demonstrated in the video clip for this activity. Where does the red dye enter the crab's body? Where do you think the water flow leads to? How does the animal create this water current?

On your specimen, identify where water is taken in (the **incurrent aperture**), and where the excurrent flow exits the body (the **excurrent aperture**).

7. To examine the structure that produces this water current, grasp the base of the 2nd maxilla firmly with your forceps and pull toward you. This should remove the maxilla, along with its attached **gill bailer,** from the excurrent aperture.

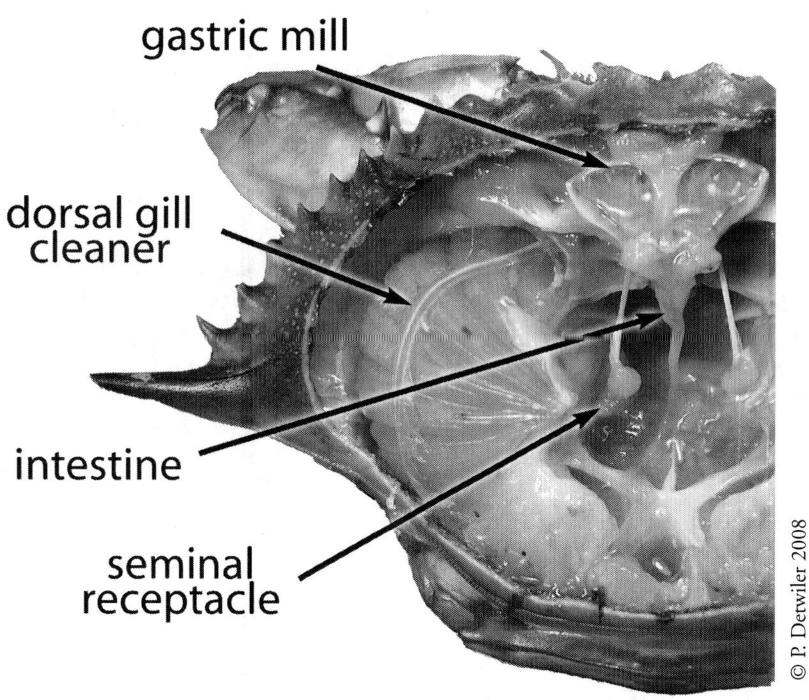

Figure 9. Internal organs of the female blue crab. Gonad, digestive gland, and gill chamber membrane have been removed to show underlying structures.

5. Locomotion and Muscles

1. Observe a living crab as it walks and describe its locomotion. Can crabs walk forward or backward?

2. Up to this point, you have seen the internal organ systems, but where is the crabmeat? Turn your attention to the pointed **endoskeleton,** upon which the gills rest. The crab locomotes with its walking legs, so logically, the muscles that control them must be found near the attachment of the legs to the body.

3. Cut a slice through the side wall of the endoskeleton to expose the muscles of the pereopods, the bulk of the crabmeat in this species.

   What body part of the Alaskan king crab provides the main source of crabmeat for human consumption?

4. Remove a cheliped and crack it open to expose the muscles controlling the pincher claw.

5. With a scalpel, scrape away the muscle tissue to reveal a central, broad plate. The plate attaches to the pincher at one end, and the claw muscles attach to the plate at the other end.

6. Pull the plate toward you to demonstrate the action of the claw muscle. What happens when the claw muscle contracts?

EXERCISE 4: COMPARISON OF CRUSTACEANS

Now examine the shrimp or crayfish your instructor has provided, and compare its external anatomy to the crab's. Record your observations on Question 2 of the following review sheet.

You have now completed this activity. Clean up your workstation according to your teacher's instructions, wash your hands thoroughly, and complete the following review sheet.

Name: ____________________
Date: ____________________
Section: ____________________

REVIEW:

1. **DVD:** After viewing the video clips on crustacean diversity, compare and contrast the modifications in body shape of several of the various decapod species shown. Hypothesize how the various modifications provide a survival advantage to each animal in its habitat.

2. Compare the anatomical modifications of a shrimp and crab. If you decide they are similar in certain characteristics and functions, place a check mark in the column. If you see differences, record it, as indicated in the table.

Crab feature:	Similar to shrimp?	If not similar, then describe the difference in anatomy and/or function:
Exoskeleton		
Cephalothorax		
Rostrum		
Eyes		
Antennae		
Maxillipeds		
Maxillae		
Mandible		
1st Pereopod		
Walking Legs (pereopods 2–5)		
Abdomen		
Pleopods		
Telson		

3. Match the terms with their correct descriptions.

1.	______ Autotomy	a.	Paired structures that decapods use for biting food
2.	______ Cephalothorax	b.	Modified appendages in male sea spiders
3.	______ Chelicerae	c.	The anterior body region of chelicerates
4.	______ Chelipeds	d.	Female decapods lay eggs through these abdominal openings
5.	______ Chitin	e.	Thin extensions of chitin used as respiratory surfaces in decapods
6.	______ Endoskeleton	f.	Structures that "chew" food for the horseshoe crab
7.	______ Exoskeleton	g.	An individual unit of light detection in the compound eye
8.	______ Gill bailer	h.	The act of cutting off a pereopod for predator evasion
9.	______ Gill cleaners	i.	The terminal segment on a decapod
10.	______ Gnathobases	j.	Abdominal appendages of decapods, modified for internal fertilization
11.	______ Gonopods	k.	This structural material constitutes the arthropod exoskeleton
12.	______ Gonopores	l.	The structures used by the horseshoe crab for forward locomotion
13.	______ Lamellae	m.	These paired structures produce spermatophores in male decapods
14.	______ Mandible	n.	Abdominal appendages of decapods
15.	______ Maxillipeds	o.	The first pair of pereopods on a crab
16.	______ Molting	p.	This body region houses the book gills of the horseshoe crab
17.	______ Neuston	q.	The anterior body region of crustaceans
18.	______ Ommatidium	r.	These structures are attached to maxillipeds and assist in gill ventilation
19.	______ Ophisthosoma	s.	The food-manipulating appendages of crustaceans
20.	______ Ovigerous legs	t.	The periodic shedding of the arthropod exoskeleton to allow body growth
21.	______ Pleopods	u.	This structure supports the crab gills and houses the pereopod muscles
22.	______ Proboscis	v.	Food-manipulating structures on the horseshoe crab
23.	______ Prosoma	w.	A jointed body covering, unique to the phylum Arthropoda
24.	______ Pusher legs	x.	Marine water striders are grouped in this category, based on their habitat
25.	______ Telson	y.	Pycnogonids use this sucking structure to ingest food
26.	______ Vas deferens	z.	A structure in the excurrent aperture of crabs that creates a respiratory current

WEB EXPLORATIONS FOR FURTHER STUDY:

Explore these relevant Websites to stimulate interest in current marine arthropod research and issues.

Monterey Bay Aquarium Seafood Watch http://www.mbayaq.org/cr/seafoodwatch.asp
What kind of crustaceans do you like to eat? Would you still choose to eat those species if they were in danger of disappearing from overharvesting, or if the marine environment was being destroyed in order to harvest those species? The Monterey Bay Aquarium began the Seafood Watch program in 1997 to raise consumer awareness about the environmental impacts of current methods of harvesting marine fish and shellfish, and to educate people so that they can have the option to buy and consume seafood collected in environmentally friendly, sustainable ways.

- **Activity:** On the main page, go to the "Browse all Seafood" link and explore the links for crab, lobster, and shrimp. What are some of the species that Monterey Bay recommends consumers avoid, and why? What are some of the environmental concerns? Which species are being overfished?
- Are there any species that you like to eat listed as species to avoid? If there are, find out what other kind of crustaceans could be substituted; i.e., which other types of crustaceans are fished in environmentally sustainable ways?
- Download a pocket guide and carry it with you to consult whenever you order seafood.

Fish Information Network http://www.globefish.org
The United Nations Food and Agriculture Organization is an international agency that works toward helping countries improve their food production. GLOBEFISH is a unit in the FAO Fisheries Department that conducts studies of international fish trade and analyzies trends in the world fish market.

- **Activity**: Visit their homepage and read the summary of the USA Market Report for crabs, lobster, or shrimp for the current or previous year. Which countries are the main suppliers to the United States of each kind of crustacean? What is the current status of the crab, lobster, and shrimp fisheries in Europe, Asia, and South America?

Maryland Sea Grant Blue Crab Information Page
http://www.mdsg.umd.edu/issues/chesapeake/blue_crabs/
An all-purpose resource on current research and management issues concerning the marine and estuarine organisms of the Chesapeake Bay.

- **Activity:** Explore the site to learn why the blue crab population is dwindling. Follow the link to the "Chesapeake Quarterly" magazine. What is being done to offset some the factors thought to be responsible for the decline? What can average citizens do to both aid conservation and reduce their impact on the Chesapeake Bay ecosystem?

Mid-Atlantic Sea Grant Network and NOAA Horseshoe Crab Information Site.
http://www.ceoe.udel.edu/horseshoecrab/
This site provides the public with current ecological, medical, and zoological research about the Atlantic horseshoe crab.

- **Activity:** How are horseshoe crabs important to humans and birds? Investigate the 'Shorebird Connection' and 'Human Use' links to find out. Click on the 'Fisheries Management' link and discover the current status of the horseshoe crab population on the Mid-Atlantic Coast. Report on your findings. Why did the horseshoe crab population decline, and what is being done to counteract the threats to the population?

Name: ____________________
Date: ____________________
Section: ____________________

Topic: ____

WEB EXPLORATION REPORT

LABORATORY 14

Phylum Echinodermata—Sea Urchin Biology

INTRODUCTION:

Echinoderms (Greek: *echinos* = hedgehog; *derm* = skin; "spiny-skinned" animals) are exclusively marine; no species live in freshwater. These animals are commonly found on rocky shorelines throughout the world, and are also abundant on the ocean floor, forming up to 80% of the **biomass** of the deepsea epifauna in some areas. The five classes in the phylum are listed below, with the approximate number of species recorded for each class:

A. Echinoidea: sea urchin and sand dollars (950)
B. Asteroidea: sea stars (2,000)
C. Ophiuroidea: brittle stars and basket stars (2,000)
D. Holothuroidea: sea cucumbers (1,100)
E. Crinoidea: sea lilies and feather stars (600)

Figure 1. Diversity of echinoderms, showing typical representatives in each class.

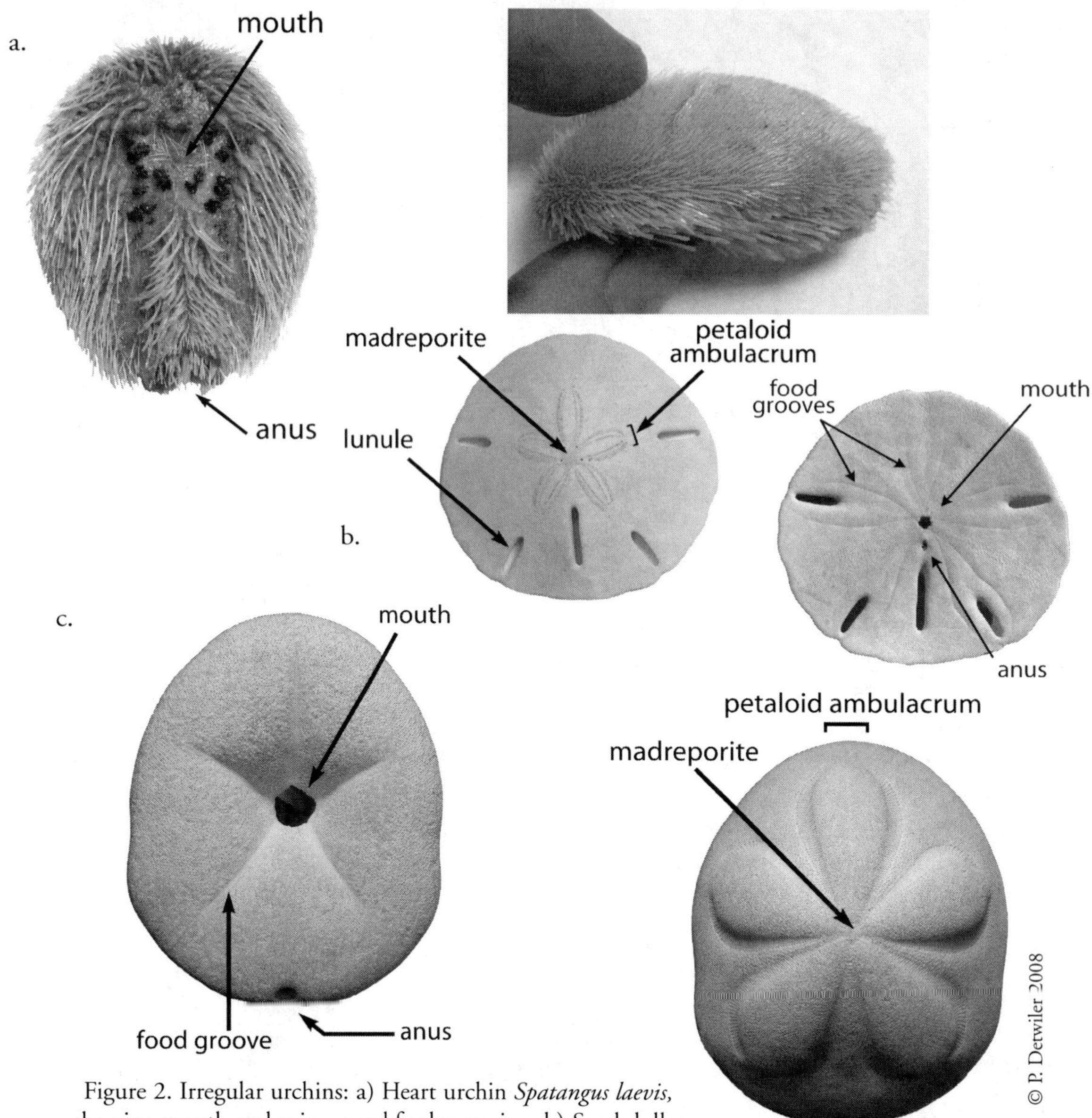

Figure 2. Irregular urchins: a) Heart urchin *Spatangus laevis,* showing mouth and spines used for burrowing; b) Sand dollar *Mellita quinquiesperforata*, 35 mm; c) Caribbean sea biscuit *Clypeaster rosaceus,* 15 cm.

EXERCISE I. SEA URCHIN ANATOMY

Sea urchins and sand dollars are **echinoids**, members of the class Echinoidea. Species in this taxon characteristically have a **benthic** lifestyle and are found throughout the world's oceans—from the crashing surf of the rocky intertidal zone to abyssal depths over 4 kilometers below the ocean surface. Echinoid body shapes vary depending on the particular lifestyle of each species. Round urchins are called **regular** urchins and are **epifaunal** organisms, nestling in crevices or crawling over the sea bottom in search of food. Other species (such as heart urchins, Fig. 2a) are **infaunal deposit feeders**, burrowing through sandy bottoms and ingesting sediments, much like earthworms do. Because their body shapes are less rounded, they are termed **irregular** urchins. Sand dollars and sea biscuits (Figs. 2b, c) are epifaunal irregular urchins. Food is gathered and passed along grooves to the central mouth. While these types may differ in shape, all urchins have these similar anatomical features:

- a **test** (a calcareous skeleton made of numerous interlocking skeletal plates)
- **spines** of varying length
- a body with **pentamerous** symmetry (based on patterns of 5)
- a **water vascular system** that uses water-filled tubes and water pressure for numerous functions, such as anchoring the animal to the **substrate**, capturing food, sensing the environment, gas exchange, and locomotion.

Recognize the pentamerous body regions of the sea urchin as you study its anatomy today.

In shallow water environments, herbivorous sea urchins exert a strong ecological effect on the underwater community because they are major **grazers** and can cause significant changes in the abundance of seaweeds. Each coast has its key echinoid species. On the rocky Northeast Coast, the green urchin *Strongylocentrotus droebachiensis* and purple urchin *Arbacia punctulata* are prevalent. The seagrass urchin *Lytechinus variegatus* is found on the Southeast Atlantic and Gulf Coasts. Purple and red urchins, *Strongylocentrotus purpuratus* and *S. franciscanus,* are common on rocky reefs along the Pacific Coast. The latter species is heavily fished for its **roe**, sold domestically as a delicacy to Japanese restaurants in the United States, and also exported overseas to Asian markets. If you have ever eaten the type of sushi called *uni*, then you have eaten the reproductive organs of this species!

MATERIALS:

- Forceps
- Live sea urchins
- Cleaned tests of sea urchins
- Bowls of chilled, clean seawater
- Dried specimens of sand dollars and other echinoids
- Compound microscope, slides, coverslips, dissecting microscope

DVD: Watch the video clip showing how *uni* is commercially prepared at a seafood processing plant in San Diego, California.

I. EXTERNAL ANATOMY

1. Examine a live sea urchin within an aquarium tank or in a container of seawater. The first thing you should notice is the movement of the urchin's spines and the numerous long, sucker-tipped structures called **tube feet.** The animal uses its tube feet to grab food and pass it to its mouth, and some species use their tube feet to gather bits of shell, algae, and rock to cover their tests to reduce drag from surging water and also to help camouflage themselves against predators.

   DVD: This procedure is demonstrated in the video clip.

   Additionally, tube feet are used to attach the animal to the substratum, so your urchin has most likely attached itself to the sides of the tank or container. If you roughly pry the animal off or pull it forcefully from the surface, the tube feet will tear off. Though tube feet regenerate eventually, biologists respect living things and never intentionally harm animals unnecessarily, so you must learn the proper way to detach an urchin from a surface without harming it.

   a. Grasp the urchin gently, carefully working your fingers between the spines down to the test. As the spines move out of the way from under your fingers, increase your grip slowly so that your fingers come in contact with the test. Now, with a gentle motion, turn the urchin as you would a knob or dial, twisting to the left, then right in small increments. As you feel the urchin loosening its attachment, increase the amount of twisting to larger increments. Be patient—at some point, the urchin will detach itself with little to no loss of tube feet.

b. Place the urchin into a glass bowl, cover it with seawater, and take it to your workstation for further examination.

Remember, all marine invertebrates must obtain oxygen directly from seawater through the process of diffusion through their body tissues. For every minute that your urchin is out of seawater and exposed to air, immerse it in seawater for about 15 seconds.

c. Examine the shape of its test. Is it completely round like a beach ball, or is it slightly flattened on one surface?

2. Look up through the bottom of the glass bowl to see the underside of the urchin's body. Watch the center of the surface for about a minute. Do you see any pointed white structures moving?

The white tips are the teeth of a unique feeding structure called **Aristotle's lantern** (Fig. 3). This structure is composed of a complex arrangement of teeth, jaws, numerous muscles, ligaments, and skeletal pieces that work together to open and close the jaws and move the lantern sideways, up, down, and around so the animal can bite into food. Grazing species scrape algae off rocks with their teeth, and some urchins even use their teeth and spines to bore holes into sedimentary rocks, creating a protective nook in which to shelter themselves.

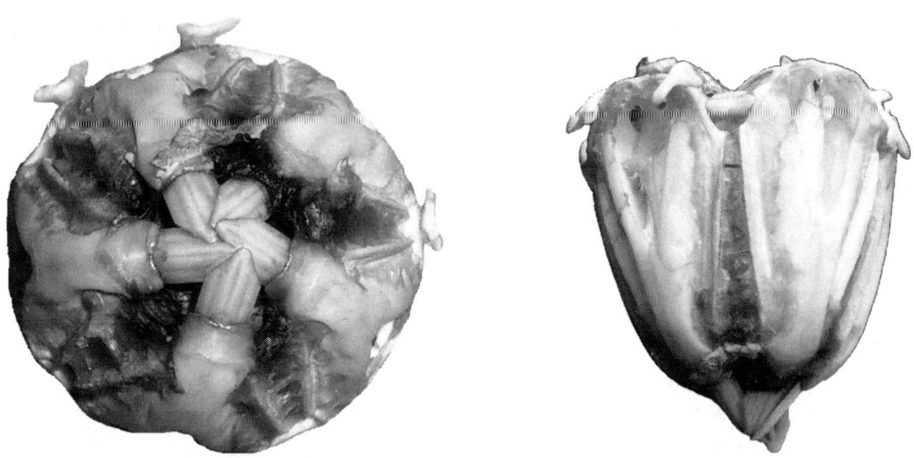

Figure 3. Aristotle's lantern, the feeding apparatus of sea urchins. Each of five jaws houses a long tooth that grows constantly as its tip wears down.

a. Is the mouth region pentamerous?

b. What do you notice about the mouth that suggests this to you?

Because the mouth is found on this bottom surface, the flattened side of the urchin's body is called the **oral surface**. The surface of the test on the opposite side of the mouth is called the **aboral surface** (*ab*: against/opposite).

3. What specialized body structures on a fish are responsible for obtaining dissolved oxygen from water?

   Do you think the sea urchin, a large animal, would also need to have similar gas exchange structures?

   What other thin tissue structures on the sea urchin's body would provide additional surfaces for diffusion of dissolved gases (carbon dioxide and oxygen)?

4. Look at the cleaned test of a non-living urchin under the dissecting microscope. The rows of pores are where the tube feet emerge. Each double row of pore-lined skeletal plates is called an **ambulacral region**. How many are there?

   The other areas with no pores are called the **interambulacral regions**. Compare these areas on a cleaned test to the same areas on the test of your living urchin. What structures are present on the interambulacral regions of a living urchin?

   a. Notice the spines—are all the spines the same length?

   b. Gently touch some spines on a living urchin. Are the spines moveable?

   c. Use forceps to twist off a small spine from the test of a dried urchin specimen, then examine it under a dissecting microscope. The spine consists of a long **shaft** and a **collar** that has a socket indentation that fits it onto a **tubercle** on the test. Notice the surface texture of the spine. Draw the spine below, labeling its parts.

Sea urchin spines come in an assortment of styles. Tropical urchins in the genus *Diadema* have spines with backwardly pointing barbs that are quite difficult to remove from unlucky divers who have bumped into them! In contrast, the kaupali or Hawaiian helmet urchin (Fig. 4), an intertidal urchin found on exposed rocks, has quite harmless spines. How are its spines modified for living in its habitat?

Figure 4. The helmet urchin (*Colobocentrotus atratus*), a common intertidal echinoid found on exposed, wave-struck rocks. Ala Moana Park, Oahu, Hawaii.

5. Using the dissecting microscope, scan the aboral surface of your living urchin. Do you see tiny structures near the surface of the test that look like little pinchers on stalks? These are **pedicellariae**. Are there different kinds? Are any of them opening and closing their jaws? Pluck one of each kind with your forceps and make a wet mount slide of each. Examine under the microscope at high power. How many jaws are there? Draw the pedicellariae below.

   What do you think these structures are used for?

   If you have a live seastar available, lay it upon the urchin so that the seastar's arm is on the urchin's oral surface. Do the pedicellariae react? Describe.

6. What structures are used for locomotion and crawling? To find out, turn your urchin upside down (oral side facing the ceiling) in its bowl of seawater and watch how it turns itself over. Which structures are more useful in this case?

If your urchin does not use its spines for crawling, what is the most probable function of its spines?

7. Now turn your attention to the aboral surface of dried urchin test. At the top, in the center of the test, is the **periproct** area (*peri*: around, *proct*: anus/rectum). When an urchin is alive, the anus is slightly off-center in the periproct area, and concealed by spines. The anus is surrounded by 5 plates that differ in appearance from all the other skeletal plates on the test. The most obvious and largest plate is usually a lighter color than the others; this is called the **madreporite** or **sieve plate**. The animal takes in seawater through this structure and circulates it throughout its water vascular system, operating the tube feet.

8. So far we have seen structures used for the processes of feeding and elimination, respiration (absorbing oxygen from the water), locomotion, and defense. What is another key life process common to all animals?

Notice that each of the 5 periproct plates bears a large pore. These pore-bearing plates are the **genital plates**, which allow a passage from the inside of the urchin to the external environment. What do you think is released out of these pores into the surrounding seawater?

II. INTERNAL ANATOMY

Study the demonstration dissection and identify the structures shown in Fig. 5.

1. Follow the digestive tract from the mouth to the anus.
2. Observe the gonads and their connection to the genital plates.
3. Follow the water vascular system from the madreporite to the ring canal to the radial canals and tube feet.

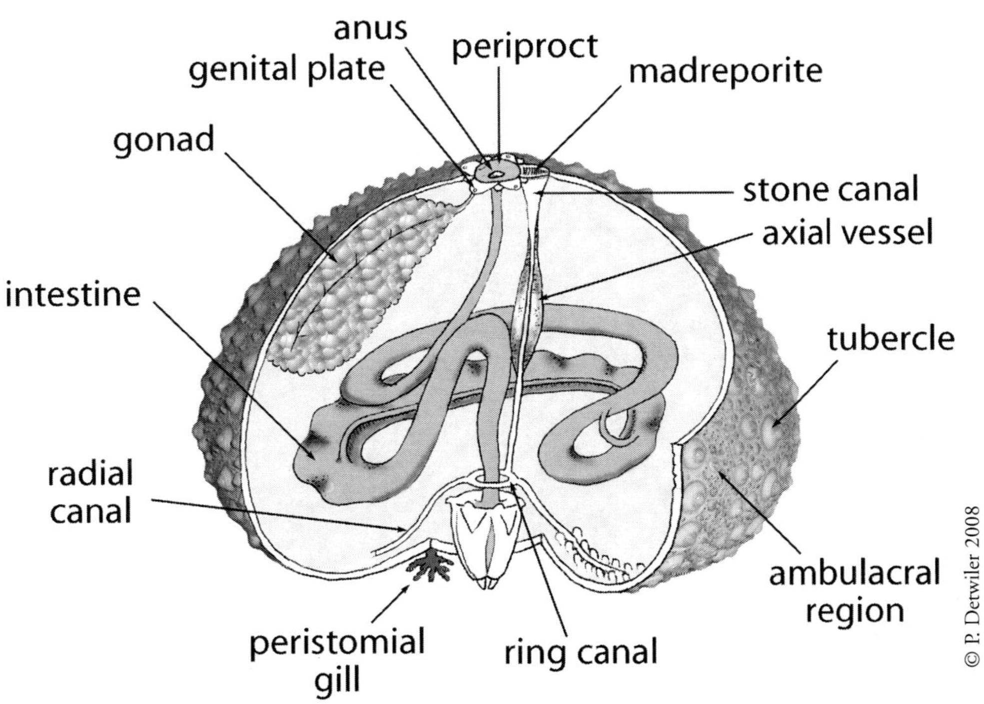

Figure 5. Internal organs of a sea urchin.

The **water vascular system** is filled with seawater that enters through the pores of the madreporite and is conducted down the **stone canal** to the **ring canal** encircling Aristotle's lantern. The water is then pumped into 5 **radial canals** attached to the inner surface of the test opposite each row of tube feet, and the canals extend up the inside curved wall of the test to the aboral surface. They are connected to the double rows of tube feet. Each radial canal emerges through a pore on an **ocular plate** and looks like a tentacle. On the stone canal is a large **axial vessel**, which is associated with excretion and may have other functions as well.

If time permits, proceed with the next exercise: Fertilization in the Sea Urchin. Otherwise, complete the worksheet on the following page.

Name: ____________
Date: ____________
Section: ____________

REVIEW:

Label the structures that you have learned today on Fig. 6.

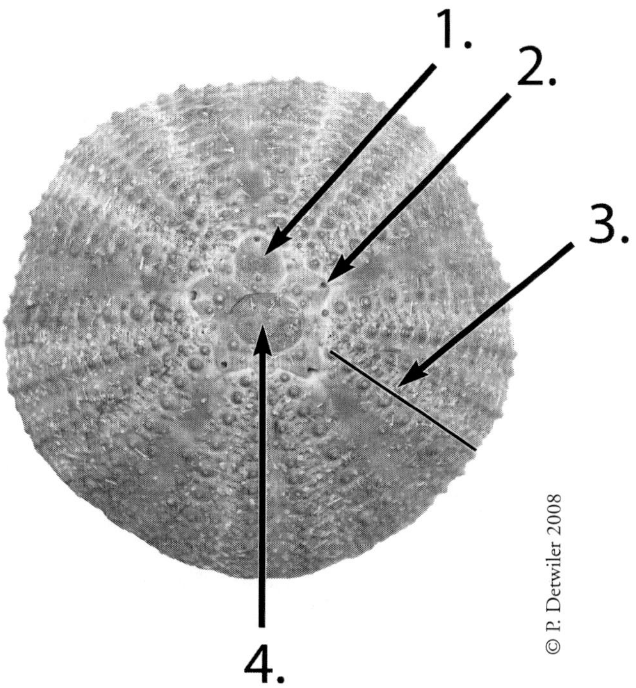

Figure 6. General features of sea urchin anatomy.

Complete the table.

Structure:	Function:
Aristotle's lantern	
Pedicellariae	
Madreporite	
Gills	
Anus	
Spines	
Genital plates	
Tube feet	

Use the words from the word bank to complete the paragraph. Not all words are used, but some words are used more than once.

Unlike our human species, which has________________________________, or 2-sided symmetry in our external appearance, the sea urchin has ________________________________, (5-sided) symmetry, which can be seen in the repetition of body parts in patterns of 5s. Two structures on the sea urchin that show this symmetry are the ________________________________ and ________________________________.

The urchin crawls by primarily using ________________________________, found on the ________________________________ surface of the animal. The tube feet are part of the ________________________________ system, a unique adaptation of echinoderms in which the animal circulates seawater in and out of its tube feet to extend and retract them. The urchin consumes food by first using its ________________________________ to transfer the food item to its mouth. Then the food is ingested through the structure called ________________________________. Any indigestible material is eliminated through the ________________________________, which is located on the ________________________________ surface of the animal. Sea urchins, like most marine invertebrates, release these two kinds of gametes (________________________________ and ________________________________) into the surrounding seawater through holes in the ________________________________, so, therefore, fertilization of these gametes occurs ________________________________. A sea urchin larva then spends several months in a ________________________________ lifestyle, drifting in the water column until it undergoes metamorphosis and changes into a tiny urchin that sinks down to the bottom, finally becoming a member of the ________________________________ community.

Aboral	Ambulacral regions	Anus	Aristotle's lantern	Bilateral	Benthic
Eggs	Externally	Genital plates	Gills	Internally	Oral
Pentamerous	Planktonic	Radial	Sperm	Teeth	Tube feet
Water vascular					

EXERCISE 2: FERTILIZATION IN THE SEA URCHIN

INTRODUCTION:

Sea urchins are ideal animals to use for studying the processes of fertilization and the subsequent early stages of embryonic development, because the pattern of cell division in echinoderms is similar to the pattern followed by members of the phylum Chordata (including humans). In the process of fertilization, **haploid** egg and sperm cells meet in the water column to form a **diploid** fertilized egg, or **zygote**. Countless studies of cell physiology and embryology have been conducted using sea urchins as model systems because they are easy to collect and handle. Marine biologists can easily induce spawning in sea urchins, and each female typically releases thousands of eggs that then can be used in medical or toxicological research.

You will obtain **gametes** (eggs and sperm) from live sea urchins that will be chemically induced to **spawn** (release their gametes out into the water column). Urchins will be injected through the **peristomial** (*peri*: around, *stom*: mouth) **membrane** with a solution of potassium chloride (KCL). You will combine the gametes, then immediately observe the initial events of fertilization under the microscope. Simultaneously, your instructor will fertilize a batch of eggs for the class to study. Throughout the remainder of the lab, you will periodically examine samples of those fertilized eggs to see later stages of cell division.

Keep in mind that what you observe is very similar to stages of fertilization and development in mammals. Sea urchin egg and sperm cells are similar in size to human eggs and sperm. When you were conceived, your first 24 hours of development proceeded much in the way you will see the sea urchin eggs developing today.

MATERIALS:

- Bowls of chilled, clean seawater
- Sea urchin gametes in labeled containers of seawater
- Compound microscope, slides, coverslips, dissecting microscope
- Two sets of disposable pipettes (one set labeled 'sperm' and the other 'eggs')

PROCEDURE:

1. First, prepare 2 depression slides, one containing a few drops of unfertilized eggs (ova), and the other a few drops of sperm suspension. Make sure to use the appropriate pipette to avoid cross-contaminating the eggs and sperm. Examine a drop of the diluted sperm solution. Sea urchin sperm cells are much smaller than the eggs, so you may have to adjust the iris diaphragm to create contrast to see the sperm more clearly. The male urchin's DNA is carried in the head of the sperm, and the sperm cell is **flagellated** (meaning it has a **flagellum**, a long whip-like structure that propels the sperm forward through the water). Draw a few sperm cells below.

2. Observe the eggs under the microscope at high power magnification. Notice that the yolk contains food for the developing embryo until it develops into a **pluteus larva**, the stage when it is able to feed on phytoplankton in the water column. The egg is surrounded by a clear **vitelline membrane**. Draw an unfertilized egg below:

Keep your depression slide of unfertilized eggs in sharp focus at medium power. Carefully remove the slide from the microscope stage by sliding it out toward you, but DO NOT change the objectives or touch the focus knobs. You want to be able to slide your next preparation of fertilized eggs onto the stage and have the slide be in focus so you do not miss the initial events occurring immediately following fertilization.

3. These next steps have to happen quickly, but work carefully.

   a. Place a drop of fresh eggs into a new depression slide; [Note the time, and set time = 0 min.] then add a drop of the diluted sperm suspension.
   b. Cover the slide with a coverslip, and carefully slide the preparation into position under the microscope, being careful not to touch the objective with the slide. Very soon after the eggs and sperm are mixed, sperm will attach to the egg, and when one penetrates the vitelline membrane, a series of chemical reactions immediately ensue, causing a **fertilization membrane** to form around the egg.
   c. Record the time when fertilization membranes begin to appear: ________________

These fertilized eggs are now called **zygotes**. The fertilization membrane forms a tough barrier around the zygote to prevent other sperm from entering. Why is this important? What would happen to the genetic composition of the zygote if multiple sperm cells fertilized the egg at the same time?

Draw and label a newly fertilized egg below in space A:

A: t = 0 min.	B: t = 30 min.	C: t = 60 min.

4. The first cell division occurs about 30 minutes after the nucleus becomes visible, or about 60 minutes after fertilization. Each successive cell division or cleavage occurs about every 20 minutes thereafter. Get samples of fertilized eggs from your instructor at the indicated times.
    a. When about 30 minutes have elapsed (t = 30 min., a half hour after fertilization has occurred), the **haploid** set of chromosomes from the sperm combine with the haploid set of chromosomes from the egg to form a **diploid** nucleus. Do you see examples of this? Draw this in space B. When you are finished drawing, rinse the eggs into the waste container of seawater.
    b. At time = 60 min., remove a few drops of fertilized eggs and make a drawing of the first cleavage in C. Rinse the eggs into the waste container of seawater.
    c. At 20-minute intervals, remove a sample of eggs, observe them, and in boxes D through F, draw the cleavage stage shown by the majority of eggs.

D: t = 80 min.

E: t = 100 min.

F: t = 120 min.

Have the zygotes increased in size as they pass through the 2-, 4-, and 8-cell stages?

5. Refer to Figure 7 for subsequent stages of sea urchin development. After 6 hours, a hollow ball of ciliated cells forms, called the **blastula**. If your instructor has specimens available at this stage, obtain some and observe under the microscope. Note that the blastulas can swim through the water. What are they swimming with? Focus at high power and adjust the iris diaphragm to see the structures responsible for the blastula's movement. These little hairlike threads are **cilia**, and they line the outside cell layer of the blastula.

Human and urchin development proceeds in different ways after the blastula appears. **Gastrulation** (the initial formation of the digestive tract) occurs about 1–2 days later in the sea urchin; during this stage, an opening appears (the **blastopore**) that ultimately becomes the animal's anus, and the primitive gut (the **archenteron**) forms. The **pluteus** larval stage arises soon after the gastrula, but you are not likely to see these stages because it is difficult to keep marine larvae alive for periods longer than a day without specialized methods of care. The pluteus larvae drift in the water column as part of the plankton community for several months, and larvae may be transported thousands of miles away from where they began life as a zygote. A miniature version of an urchin called the **rudiment** develops within the maturing pluteus body, and when the time is right, the pluteus larva undergoes metamorphosis, and the rudiment becomes the new test of a tiny urchin that settles to the bottom. From then on, the urchin is part of the benthic community.

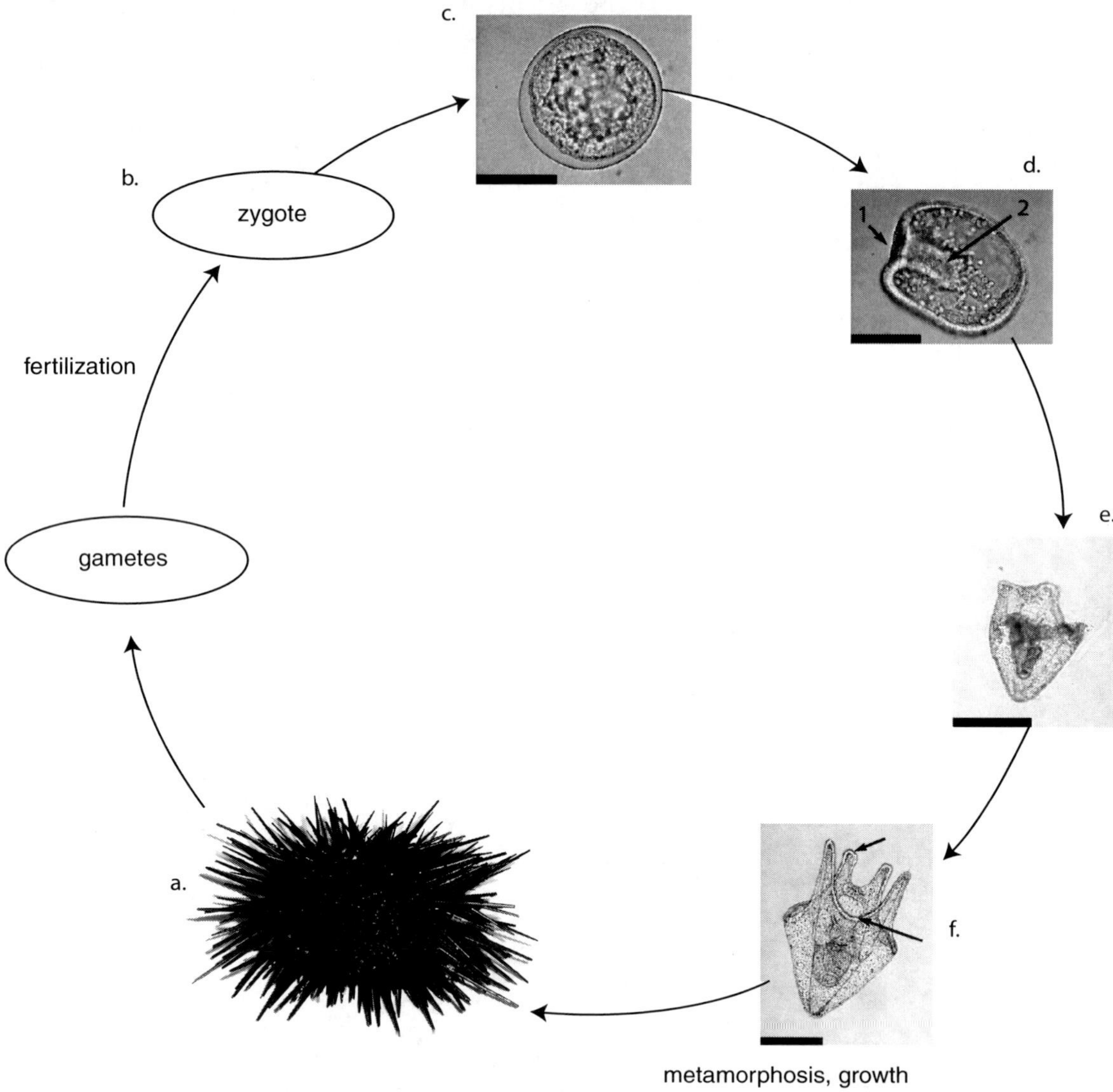

Figure 7. Developmental cycle of the sea urchin *Strongylocentrotus*: a) adult; b) zygote; c) blastula formed of a layer of ciliated cells surrounding a cavity (the blastocoel); d) gastrula showing the (1) blastopore and (2) archenteron. Scale bars are 100 µm; e) early pluteus stage; f) late pluteus stage. Arrows point to bands of tissue lined with cilia. Scale bar = 200 µm. Microscopic images courtesy of M. Russell, Villanova University.

Name: ______________
Date: ______________
Section: ______________

REVIEW QUESTIONS:

1. What visual sign indicates when an egg has been fertilized by a sperm cell?

2. What is the end result of first cleavage? Second cleavage? Sixth cleavage?

3. Describe the main difference between the blastula and gastrula stages.

4. What evidence from your observations of echinoid development (Fig. 7) suggests that radial symmetry was a **secondary development** in the evolution of echinoid characteristics?

5. Are sea urchin larvae **planktotrophic** or **lecithotrophic**? What structures on the larvae suggest an answer to this question? (Hint: See Plate 4 of Laboratory 7: Observing Plankton.)

FOR FURTHER STUDY:

Sea urchin eggs are routinely used in bioassays to test the toxicity of various chemicals. Many chemicals used in common household cleaning products are washed down drains and are unaffected by sewage treatment procedures before being released as wastewater into waterways and coastal regions. How might these chemicals affect marine invertebrate populations? Devise a controlled experiment to investigate which brands of common household cleaning products may be more environmentally-friendly to marine life (i.e., which have the least harmful effects on survival and development of sea urchin zygotes).

Name: ____________________
Date: ____________________
Section: ____________________

Topic: ____

WEB EXPLORATION REPORT

Name: ______________
Period: ______________
Date: 11/16/11

REVIEW:

1. State the functions of blood circulation.
To pump blood throughout the body of an animal.

2. What is the relationship between metabolic rate and rate of blood circulation (*positive, negative, or no relationship*)? There is a positive relationship

3. Recall the relationship between dissolved oxygen and temperature. As water increases in temperature, its ability to retain dissolved gases (*increases or decreases)* decreases. Therefore, as the summer sun warms water in a shallow body of water, a fish's need for oxygen increases in the face of the decreasing level of oxygen in the water.

4. What is the relationship between metabolic rate and ambient temperature in an ectotherm? Graph it below in Fig. A. How does the relationship differ in an endoderm? Graph your hypothesis below in Fig. B.

A: Ectotherm

Metabolic rate

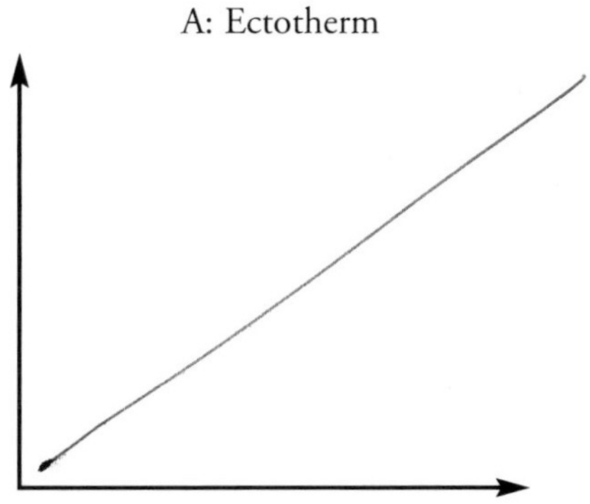

Temperature (°C)

B: Endotherm

Metabolic rate

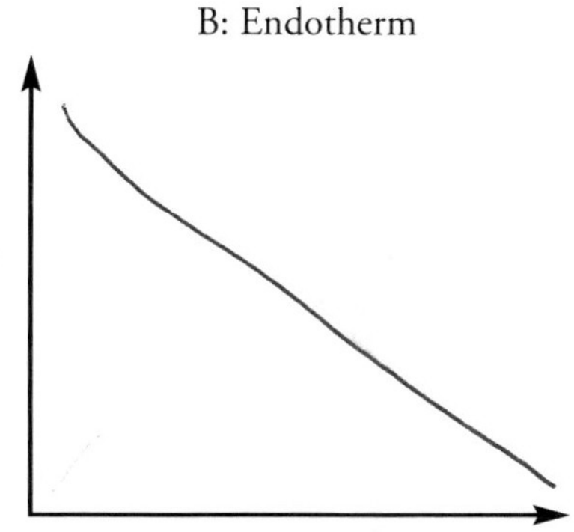

Temperature (°C)

5. Under each set of circumstances, which animal would you expect to have an overall higher metabolism? Mark your choice by checking the box; then explain why you chose your answer.

Similar-sized individuals; different habitats	
☒ 8-cm long mussel from abyssal zone (4000 m)	☑ 8-cm long mussel from intertidal zone
Why: Because it lives in a	
Same species, different sizes	
❑ 10-cm long adult pufferfish	❑ 3-cm long juvenile pufferfish
Why:	

6. How is a freshwater fish's activity affected during winter in cold climates?

The fresh water fishes metabolic rate increases to deal with the harsh environment.

7. How is the information you learned today valuable for applications in aquaculture?

It's helpful to find the adaquete temperature for raising fish.

8. Research how the activity level of cellular enzymes changes with temperature. What happens to the rate of enzyme reactions at temperature extremes? What would happen to the functioning of cells at those extremes?

During temperature extremes the reactions of enzymes becomes slower, during these periods they can become deactivated

9. During summer months at the Salton Sea (California's largest lake, located in a desert region), fish are sometimes seen gulping air at the water surface. Offer an explanation using ideas from this exercise.

The lakes salinity increases making it difficult for the fish to breathe and thrive

10. What do marine scientists believe about how global climate change and increased sea surface temperature affect reef-building corals?

The increase in climate increases the acidification and bleaching in coral reefs.

FIELD STUDY 1

Oceanography Activities on a Pier or Dock

During this period in the field, you will:

- Measure physical and chemical properties of seawater, using standard techniques.
- Collect and process seawater samples for later analyses in the laboratory.
- Collect marine plankton and compute its density.
- Record observations on marine species encountered in the field.

OVERVIEW:

You will be assigned to a group, which will begin working on one of the following exercises. Each group will perform a certain task or observe a demonstration by your instructor, then rotate to the next exercise. Follow your instructor's directions for using all equipment and tools at each station, and record your observations and data on the appended datasheet. Do not share data with other groups. When you have completed the activity at your station, check with your instructor before moving to the next station. All groups will rotate through all the stations and perform all the required tasks during the field study period. At the conclusion of the field trip, all the collected data will be consolidated for discussion of general trends and findings.

DVD: Many of these procedures are demonstrated in the DVD.

EXERCISE 1: PHYSICAL OCEANOGRAPHY—LIGHT PENETRANCE/WATER CLARITY

One branch of oceanography is concerned with the behavior of sunlight as it penetrates the water's surface. This phenomenon is quite complex and involves reflection, diffraction, refraction, scattering, absorption, and polarization. These processes are affected by particulate materials in the water column, such as sediments and plankton, as well as dissolved chemicals. In addition, light intensity fluctuates constantly throughout the day, due to variation in the angle of the sun and changes in cloud cover. The optical properties of seawater are very important for photosynthetic organisms (phytoplankton, seaweeds, sea grasses), as well as for shrimp, fish, and other **zooplanktonic** animals that migrate vertically through the water column each day in response to changing light levels.

A **Secchi disc** (Fig. 1) is a weighted, white disk used to measure water clarity to the depth the naked eye can see through water. Clarity is the opposite of turbidity. Clarity is decreased by the amount of **organic** (containing the element carbon) and **inorganic** particulate matter suspended in the water. These materials scatter light beams streaming down from the sky. Each particle of suspended matter reflects a tiny amount of light, scattering it in a random direction. For an observer at the surface to view any kind of submerged object, light beams must reflect off the object and return upward into the viewer's eyes. However, when an object is at a certain depth, all the reflected light beams are scattered sideways, and none reach the observer at the surface. When this occurs, the submerged object disappears from view (the vanishing point, or the **Secchi depth**, is reached). Thus, Secchi depth is inversely related to water turbidity.

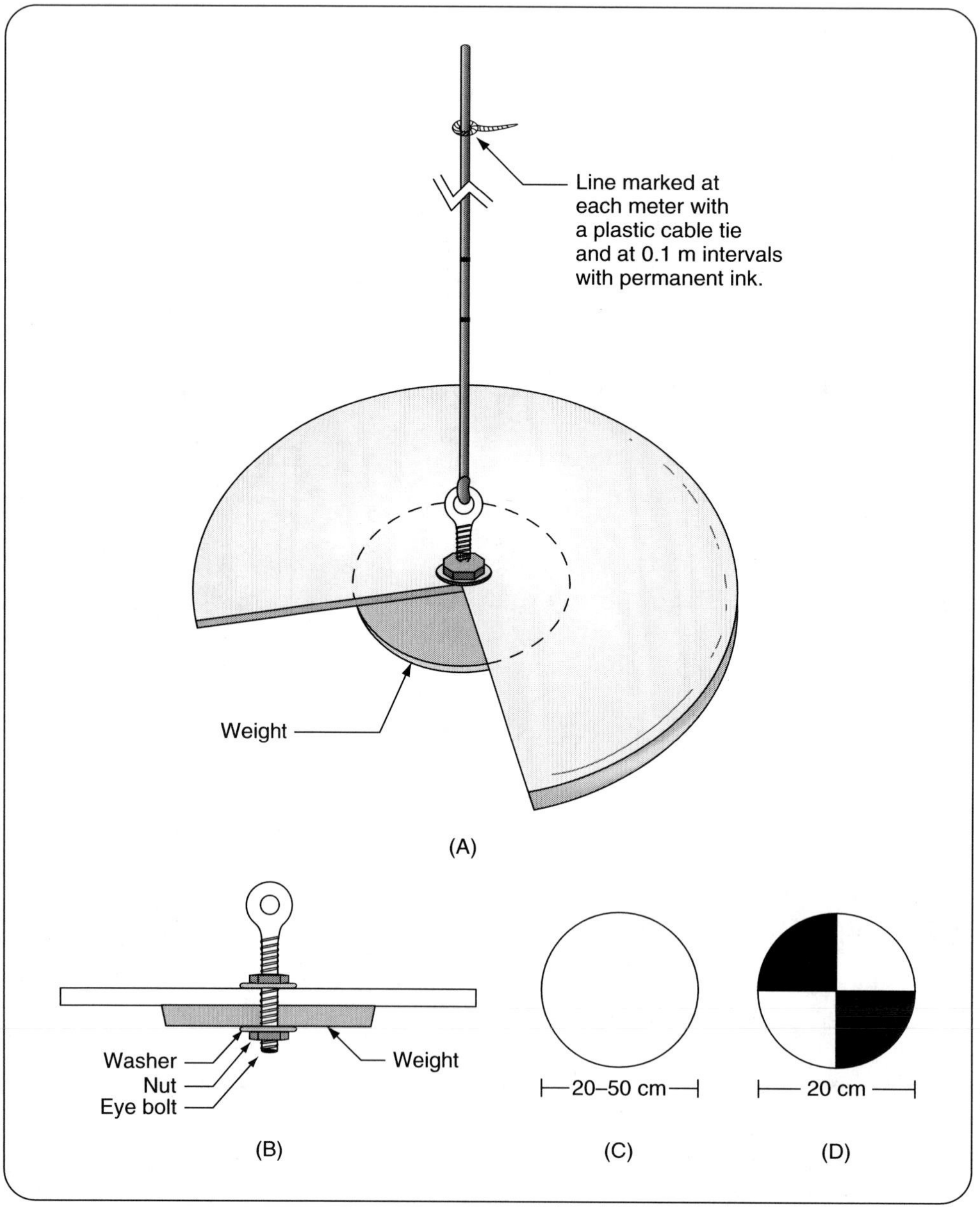

Figure 1. Construction of a Secchi disc. a) top view; b) cross-section. Discs used in oceanic sampling typically are white as in c), while freshwater Secchi discs usually have black and white quadrats (d).

The Secchi disc is lowered into water on a line that has markings to indicate the depth at which the disk is no longer visible. This simple procedure allows comparison of seasonal differences in water transparency within and between water bodies. When other oceanographic measurements such as nutrient level and plankton density are also recorded at the same location, marine scientists are able to understand meaningful relationships between these variables.

PROCEDURE:

Tie the end of the Secchi disc line onto the pier railing, using a bowline knot. Have your instructor *check the knot* before you lower the Secchi disc over the side.

1. Lower the Secchi disk into the water until it just disappears from view. If it is a sunny day, the Secchi disk should be lowered into a spot that is shaded because reflected sunlight and surface glare make it difficult to recognize the vanishing point.
2. Record the depth (to 0.1 m) where the disk is no longer visible—in other words, count the marks on the line as the disk descends below the water surface until it reaches the vanishing point.
3. Repeat Steps 1 and 2 twice more to find the average depth of the vanishing point.
4. Haul the disk up back onto the pier, and coil the line for the next group.

EXERCISE 2: COLLECTING A WATER SAMPLE FOR PHYSICAL AND CHEMICAL ANALYSIS

Several types of equipment are used to collect seawater samples from the surface, as well as from specific depths, but most sampling is conducted using the Niskin and Van Dorn types of collection devices. These tools are designed as open tubes that close at predetermined depths when a **messenger** (a sliding metal weight) is sent down the line and strikes a trigger, closing both ends of the device. The sampler remains sealed so the chemical composition of the water collected at the particular depth is not contaminated by shallower water as the device is hauled up to the surface.

PROCEDURE:

Tie the end of the water sampler line onto the pier railing, using a bowline. Have your instructor *check the knot* before you lower it over the side.

Follow your instructor's directions for operating the water sampler and for transferring the contents of the sampler into containers. Provide a water sample to the group that will perform physical and chemical measurements (Exercises 3 and 4). Keep in mind these guidelines for handling samples to be analyzed for oxygen content:

1. Slowly fill the container to overflowing by placing the hose from the sampler device in the container and allowing the water level to rise gradually, without introducing turbulence or bubbles into the water.
2. Remove the hose as it continues to drain, so that the sample bottle is completely filled, with no air space.
3. Seal the sample bottle tightly without trapping any air bubbles, or chemically stabilize the sample as your instructor advises.

EXERCISE 3: CHEMICAL OCEANOGRAPHY—DISSOLVED OXYGEN AND BIOLOGICAL OXYGEN DEMAND

One important indicator of water quality is the amount of dissolved oxygen contained within a water sample. Dissolved oxygen (DO) is a critical element in seawater because nearly every type of organism (except certain kinds of **anaerobic** bacteria) requires oxygen for its metabolism. Most aquatic organisms need a DO content of at least 4 ppm to survive.

At sea, blowing winds can form whitecaps (breaking waves), and this turbulence aerates the surface water with atmospheric oxygen. Similarly, waves arriving on the shore break at the shoreline, oxygenating the surf zone.

Cellular respiration is a process wherein sugars are consumed in the presence of oxygen, producing carbon dioxide, which is released into the environment. This reaction occurs within the mitochondria of all eukaryotic cells that derive energy from **aerobic** respiration. The opposite reaction (photosynthesis) occurs in plant and algal cells, which absorb carbon dioxide from the environment and release oxygen.
These processes are summarized by the following equations:

$$C_6H_{12}O_6 + 6O_2 \longrightarrow 6CO_2 + 6H_2O \quad \text{(Respiration)}$$

$$6CO_2 + 6H_2O \longrightarrow C_6H_{12}O_6 + 6O_2 \quad \text{(Photosynthesis)}$$

In addition to cellular respiration, several other factors may decrease the amount of dissolved oxygen in a water body. **Water temperature** affects the capacity of water to hold dissolved oxygen. As water temperature increases, the amount of oxygen able to remain dissolved in water rapidly decreases. Thus, the level of DO in a particular water body usually fluctuates with the changing seasons.

The **concentration of dissolved salts** in a water body also affects the DO level; the more saline a water sample is, the less oxygen is able to stay dissolved in it.

The decomposition of **organic matter** rapidly depletes DO in water. If all the oxygen is removed from a water sample, it is said to be **anoxic**.

Finally, the amount of nutrients contained in a body of water (its **trophic state**) influences the amount of oxygen that can remain dissolved within it. When a body of water is **eutrophic**, nutrient levels are high, stimulating the growth of plants and algae. Coastal waters are usually greenish because of this phytoplankton growth. Under extremely eutrophic conditions, such as within polluted coastal embayments and inland lakes, phytoplankton blooms can decrease water clarity to Secchi depths of < 1 m. At such high density, plankton populations often "crash," and their subsequent decomposition strips the water column of oxygen within hours, causing fish kills. Thus, it is not uncommon for dissolved oxygen levels in shallow-water eutrophic environments to fluctuate widely over short-term periods.

In contrast, an **oligotrophic** water body has high levels of dissolved oxygen, but low levels of dissolved nutrients. Consequently, phytoplankton density remains low and the water is extremely clear. This is typical of **pelagic**, open ocean water ("blue water"), far offshore from coastal influences. Secchi depths in the Sargasso Sea (located at the center of the North Atlantic Ocean) have been recorded as deep as 66 m. In these marine environments, dissolved oxygen content in surface waters remains relatively constant.

PROCEDURE:

Obtain a sample of ocean water from the water-sampling group. Fill a glass bottle with seawater to overflowing, but do not agitate the sample or trap any air bubbles in the bottle. Use any of the methods below to determine dissolved oxygen content.

1. Dissolved oxygen content can be measured with an oxygen probe. If the cable is long enough, the probe can be lowered to various depths in the water column, or it can be inserted into a water sample collected with the water sampler.
2. DO can also be approximated by using a colorimetric test. Consult the instructions for any chemical testing kits that your instructor provides.
3. The Winkler titration can also be used to determine dissolved oxygen content, though it is inconvenient to perform in the field. Usually a water sample is chemically stabilized in the field at the time of collection, and the titration occurs later, back in the laboratory.

BIOLOGICAL OXYGEN DEMAND:

In the process of breaking down biodegradable, organic materials into simpler chemicals, bacteria and other microorganisms in seawater remove large quantities of dissolved oxygen from the water column. Biological Oxygen Demand (BOD) is a method used to determine the amount of oxygen consumed in a water sample held in the dark at a constant temperature (20° C) for a certain number of days. Testing for BOD provides information on the quantity of pollution/biodegradable wastes in a given water body, while oxygen content tests report on the conditions at the moment the test was made.

PROCEDURE:

To measure the 7-day BOD, fill a clean large jar about 3/4 full with the sample water, and record salinity and water temperature. Cover the jar with its lid, and shake it vigorously for 1 minute. Determine the amount of dissolved oxygen in this water, using standard oxygen-testing procedures. Record this value on your datasheet. This value, in theory, should be the maximum amount of dissolved oxygen able to be held in seawater at the salinity and temperature that you measured.

1. Pour this fully-aerated sample water into a glass-stoppered bottle, making sure the bottle is filled to overflowing and that there are no trapped air bubbles. Ideally, samples for this analysis should be collected in 300-ml glass-stoppered BOD bottles; these are specifically made for this purpose.
2. Cap the bottle and wrap the stopper with Parafilm® to keep the cap from being dislodged during transport.
3. Wrap the bottle completely with aluminum foil. You will take this back to the lab and place the bottle in a dark place for 7 days.
4. At day 7, determine oxygen content, using the same procedure you used at the field site. Subtract the 7-day oxygen value from the value recorded at the time of collection. The difference in dissolved oxygen level is the 7-day BOD for the sample. Record this value on your datasheet.

EXERCISE 4: CHEMICAL OCEANOGRAPHY—PH TESTING, SALINITY, DENSITY, AND NUTRIENTS

The pH scale measures the amount of hydrogen ions in a water sample. Numbers on the pH scale range from 0 to 14, with distilled water having a neutral pH value of 7; pH values below 7 are typical of acidic solutions, while alkaline (also called **basic**) solutions have pH values between 8 and 14. Sulfur and nitrous oxides, chemicals produced by industrial emissions and noxious components of smog, combine with water vapor in the atmosphere and form sulphuric and nitric acid, which enter the water cycle through rainfall. In some parts of Northeastern United States, acid rain is a serious environmental problem for aquatic life in freshwater lakes; many of these organisms cannot tolerate fluctuations in pH due to acid rain.

Nitrates and phosphorus are nutrients for photosynthetic organisms, and essentially act as fertilizers. In general, if levels of these nutrients are high, phytoplankton reproduce and grow rapidly in a phenomenon called a **plankton bloom**, which may turn the ocean water various shades of green, brown, or even red, depending on which species of phytoplankton are most abundant in the bloom.

PROCEDURE:

The pH of the water is tested using standard pH test paper or a digital probe.

If you use test paper:

1. Immerse the test strip in the water sample.
2. Remove it and immediately compare it with the color chart found on the test paper dispenser.

If you use a probe:

1. Immerse the electrode into the seawater sample and wait for the reading to stabilize.
2. Record the pH value on your datasheet.
3. Rinse the electrode with distilled water before storing the probe.

Use a salinity probe, refractometer, or a salinity hydrometer to determine the salinity of your sample. Your instructor will demonstrate the operation of whichever instrument is available for your use.

Nutrients are analyzed by simple colorimetric procedures that involve comparing the color of your treated seawater sample to a color chart of standards. Follow the instructions provided in the test kits to determine dissolved nitrate and phosphorous content in your sample. Record your values on your datasheet.

EXERCISE 5: BIOLOGICAL OCEANOGRAPHY—PLANKTON COLLECTION AND CALCULATION OF PLANKTON DENSITY

Use a plankton net, which collects both phyto- (plant) and zoo- (animal) plankton. Note the mesh number, which indicates the diameter of the holes in the fabric of the net. This number is given in microns (μm; 1 μm = 1/1,000 millimeter). All particles larger than the mesh holes are captured and concentrated in the **<u>cod</u> <u>end</u>** of the net (Fig. 2), while smaller particles pass through the mesh fabric.

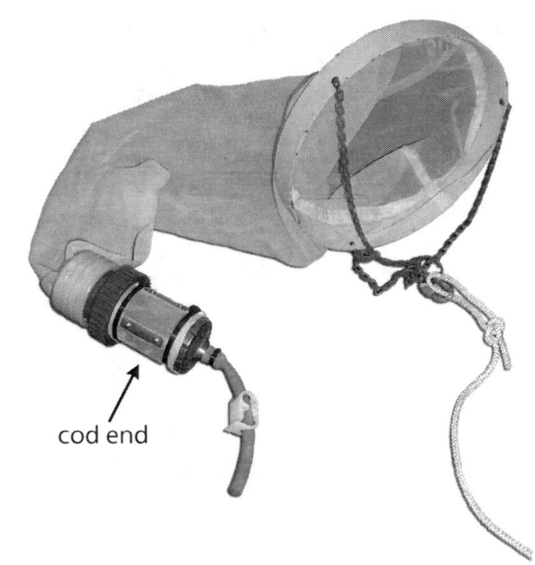

Fig. 2 Plankton Net

1. Lower the plankton net over the side of the rail, being careful not to catch the net on any snags or splinters. Submerge the net to a depth of 1 meter; then raise the net to the surface, allowing all the water to drain through. This counts as 1 vertical tow.
2. Repeat this procedure, and record the total number of tows your group performs (your group should do at least 10 tows).
3. Alternatively, if your field trip is conducted on a dock, you may walk along its length, towing the plankton net alongside in the water. Be extra careful not to catch the net on any obstructions projecting from the dock. Record the total distance you and your group members walked; this is the length of your group's horizontal tow.
4. When you have finished, haul up the net, empty the cod end container into a collection bucket, and use a squeeze bottle of seawater to rinse down the outside of the net into the bucket. It is not necessary to rinse the plankton out into the container after every tow, just at the end of your group's turn.
5. Each group will add its collected plankton sample to the bucket. At the conclusion of the exercise, your instructor will preserve the entire sample in a seawater solution of 5% buffered formalin for later analysis back in the laboratory.

Meanwhile, calculate the total amount of water (in liters) that passed through the plankton net while your group collected plankton. Use the formula for a cylinder:

Area x Length = Volume.

Area = πr^2 = (3.1416) x (radius of the plankton net in cm)^2

Length = # of vertical tows that your group conducted (1 tow = 1 m) x 100 (this step converts m to cm)

Or

= total distance (in m) that your group towed the net alongside the dock x 100 (multiplying your values by 100 converts m to cm)

Volume = __________ cm^3 Now convert to ml (1 cm^3 = 1 ml)

= __________ ml Now convert to liters (1000 ml in 1 liter)

= __________ liters

Record this value on your datasheet.

EXERCISE 6: BIOLOGICAL OCEANOGRAPHY—BENTHIC SPECIES

Bottom sediments are collected with various types of grab samplers; a common design for small samples is the petite Ponar grab (Fig. 3). Your instructor will explain the procedures to safely operate the bottom grab. Rinse the contents of the grab through a sieve, and examine the material retained on the sieve for any **infaunal** or **epifaunal** **benthic** invertebrates. If organisms are present, separate them into containers of seawater to show the rest of the group. Use local field guides to identify abundant species. If your instructor wishes, collect some of the sediment in a labeled plastic bag for later analysis back in the laboratory.

a. b. c.

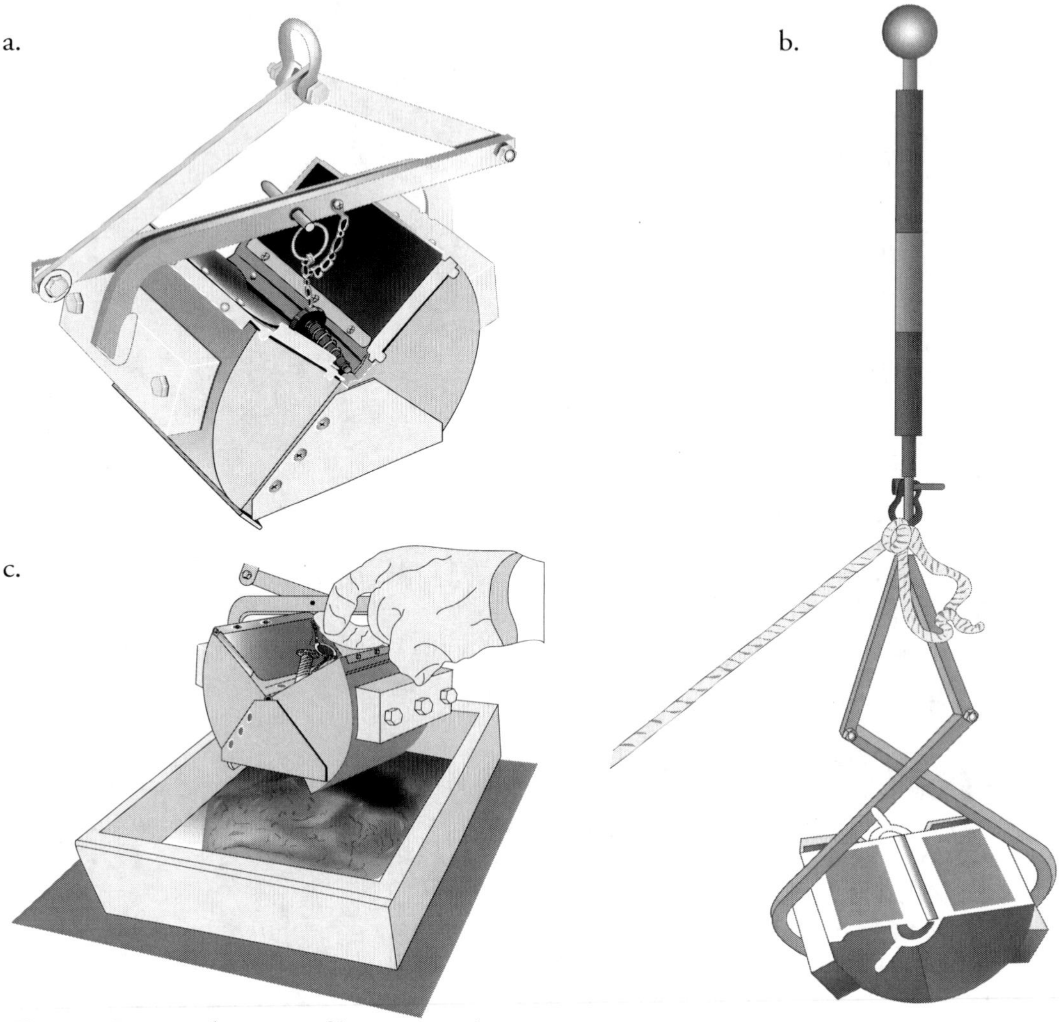

Figure 3. Ponar grab, a type of bottom sampling apparatus. a) Storage and/or transport. Safety pin keeps the unit from closing. b) Grab in closed position, containing a sample. c) Grab is opened and emptied onto a sieve. When deploying or handling the grab, wear heavy gloves for safety.

EXERCISE 7: BIOLOGICAL OCEANOGRAPHY—IDENTIFYING COASTAL SPECIES

Use local field guides to coastal birds, mammals, and fish to identify as many species of marine life as you can find during a 15-minute period as you walk up and down the pier. What species of fish are people catching? If you have binoculars, use them to watch birds on the pier or the sea surface. Scan the horizon for signs of marine mammals. Are there any sessile invertebrates attached to the pier pilings? Record all your observations and identifications in the datasheet.

DATA SHEET—OCEANOGRAPHY LAB ON PIER

Location: ______________________ Air Temperature: (°C) ________
Weather: ______________________ Date/Time: ________________
Beaufort Sea State (from Appendix E): ____________ Swells (direction, time): ______________

PHYSICAL/CHEMICAL DATA:

Secchi Disc reading (in m): __________ Class mean: __________

Water collection depth: __________
Water temperature: (°C): __________ Class mean: __________
Water density (g/ml): __________ Class mean: __________
Salinity (ppt): __________ Class mean: __________
DO of water sample (in ppm): __________ Class mean: __________
pH of sample: __________ Class mean: __________
Nitrate content (ppm): __________ Class mean: __________
Phosphorus content (ppm): __________ Class mean: __________

(A) DO of fully aerated water sample (ppm) at temperature you measured: __________
(B) DO of water sample after 7 days in dark: __________
7-day BOD (= A – B): __________ ppm
Class average: __________ ppm

Classification of 7-day BOD water sample (circle one):

1 to 2 ppm BOD	Very little biodegradable waste in water
3 to 5 ppm BOD	Relatively clean water
6 to 9 ppm BOD	Elevated levels of organic material (waste, bacteria)
$\geq$ 10 ppm BOD	Extremely high levels of organic material

PLANKTON DATA:

Diameter of net (cm): __________
Length of net (cm): __________
Mesh diameter (μm): __________
Number of vertical tows (by your group): __________
Total volume of water (liters) (your group): __________
Total number of vertical tows (all groups): __________
Total volume of water (liters) (all groups): __________
Overall plankton volume (from all groups): __________ ml
Overall plankton concentration (ml/liter): __________

Interesting birds/marine mammals/fish/invertebrates seen at pier, and their observed behaviors:

Shorebird and Seabird Species:	Number Seen:	Behaviors:

Notes:

Marine Mammal Species:	Number Seen:	Behaviors:

Notes:

List common fish species and any benthic invertebrates observed on pier pilings or present in the grab samples.

Common Name:	Scientific Name:	Phylum:

Notes:

QUESTIONS:

1. List three environmental factors that can influence the dissolved oxygen concentration of a body of water.

2. Discuss your data regarding BOD and nutrient levels. Was your sample eutrophic? What local or regional sources of pollution might potentially explain a high observed value for BOD and/or nutrients?

3. How do you think seasonal heavy rainfall affects BOD in the coastal waters offshore of major metropolitan areas?

4. Why did you wrap the 7-day BOD bottles in foil? Why was it important to keep out air bubbles?

5. Describe the relationships between Secchi depth, plankton density, and nutrient levels. What factor(s) in the marine environment may act to limit the abundance of plankton in the water column?

WEB EXPLORATIONS FOR FURTHER STUDY:

Explore these relevant websites to stimulate your interest in coastal oceanography.

NOAA National Oceanographic Data Center http://www.nodc.noaa.gov
The National Oceanographic Data Center (NODC)—comprised of the Oceanographic Data Center, National Coastal Data Development Center, World Data Center for Oceanography, and the NOAA Central Library— provides access to the world's most comprehensive sources of marine environmental data and information.

- **Activity**: Under "Project Data Sets," click on the *Coastal Water Temperature Guide,* http://www.nodc.noaa.gov/dsdt/cwtg/ and select your state. Inspect the temperature data at the station closest to the location of where your field trip occurred. Convert water temperature from Fahrenheit to Celsius, using the temperature conversions utility located at the bottom of the screen. Compare your measurement of water temperature with the NOAA average measurement for the month your field trip occurred. Were they similar? Propose an explanation for any observed differences.

- If there is a hyperlinked value in the *"Recent Temperatures"* column for the station closest to your location, click on it, and you will be directed to a page displaying recent information collected at that station. On the left side of the webpage is a menu of the available data collected by the buoy or monitoring device at the station. Click on the different menus to investigate tide predictions, meteorological observations, and other information for the station nearest you.

World Water Monitoring Day http://worldwatermonitoringday.org
This international education and outreach program builds global awareness and involvement in protecting water resources by engaging citizens in the protection of their local water resources. Participants conduct simple water monitoring tests and upload their results to the website.

- **Activity**: Visit the website and register a site in your area that you are able to monitor during the annual monitoring period (mid-September to mid-October). Inexpensive water-quality test kits are available for purchase online on the website, or consult your instructor for equipment. Share your data and experiences with others in your community, and keep a yearly record of your monitoring test results.

FIELD STUDY 2

Fouling Community

INTRODUCTION:

The **fouling community** is a special type of benthic invertebrate community that is quite different from the rocky or sandy intertidal zone, mainly because the surfaces that the fouling organisms attach to are never exposed to air. This community consists of **epifaunal** species growing on continuously submerged hard substrata, such as pier pilings, docks, and the bottoms of boats (Fig. 1). Within a marina, the fouling community is readily accessible just beneath the water surface alongside the docks—wherever undisturbed materials have been submerged in seawater for a few months or more. Well-developed communities usually contain several representatives from each major animal phylum.

Figure 1. Dockside fouling community.

The goals of this laboratory are for you to accomplish the following:

- Examine and describe the fouling organisms that are characteristic to your coastline.
- Identify the major phyla and common species found in the fouling community.
- Compare species diversity and abundance between two different fouling habitats, and propose explanations for any differences that are seen.

BACKGROUND:

Complete the table by comparing the physical and chemical factors of the fouling community environment to those of intertidal environments on the open coast.

	Rocky or Sandy Intertidal Zone	Fouling Community Habitat
Degree of exposure to wave shock: (high/low)		
Variation in water temperature		
Variation in dissolved oxygen content		
Exposure to atmosphere		
Variation in salinity level		
Relative amount of time during the day that filter feeders are able to feed		

Because these environments are so different from one another, the organisms that live in each habitat have different adaptations for survival. Are there any types of organisms from the rocky or sandy intertidal zone habitats that you would NOT expect to see in the fouling community? Why?

Your group will work together to carefully separate organisms of the fouling community into the phyla and classes to which they belong, placing them into separate containers of seawater. Each of you in the group needs to be "responsible" for different phyla and classes, meaning that you will need to organize and enumerate the total number of each type of organism in that phylum and/or class. At the end of the activity, the class will compile all the data together to show the relative abundance of each taxon in the community.

PROCEDURE:

1. Select an area that is off to one side so that you do not block other people's access along the dock. A low-use area of the marina or dock, where the surface has not been scraped or cleaned for a few seasons, is best. Try to find an area that is heavily encrusted with marine life.
2. One student holds a small bucket underwater, beneath a promising-looking section of fouling growth, while another student uses a tool such as a putty knife or spatula to pry off a 10 x 10 cm chunk of encrusting material into the bucket. Do not use the tool to hammer or break off material; instead, slide the tool along the substrate, lifting off encrusted organisms.
3. When you have detached a piece, bring the small bucket topside, and place its contents into a larger bucket. The sample should be washed several times because fouling organisms (especially mussels) tend to harbor a lot of silty sediment. To wash the sample, pour seawater into the bucket to fully cover the sample, stir, and pour out the murky water through a fine sieve or colander, retaining any small organisms trapped on the sieve. Repeat this rinsing process until the rinse water is mostly clear (this reduces the amount of mud and sediment in your sample and makes sorting easier).
4. If possible, collect a second sample from another location on the dock where there is less light exposure (such as the shady side of the dock, or even by reaching farther underneath the dock to remove a sample). If you see a submerged rope, you may be able to haul it up and compare the animals growing on a deeper part of the rope with the animals of the surface community.
5. Make a sketch of each sample, noting shapes, colors, sizes, and textures of the different organisms that you see.

How many different species are encrusting?

Do some tend to grow on a particular type of surface?

6. Using your hands, carefully separate different types of **mobile** and **sessile** organisms from the sample and place them into containers of clean seawater, keeping similar **taxa** grouped together (for example, if you have 3 different species of shrimp in your sample, place each species into a separate container and group the 3 containers close to each other, next to containers of other arthropods you may find). Use forceps and plastic spoons to gently pick up small mobile organisms.
7. As you work to separate animals and categorize them into taxa, use dichotomous keys and local field guides to help you identify the common species to the lowest taxonomic level possible. You are not responsible for the scientific (binomial) names, but you should know the common names of the most abundant organisms. Take digital pictures of any unknown organisms to aid in identification back in the laboratory. A small plastic aquarium may be used to hold specimens for photography.
8. Fill in the Fouling Community Datasheet with information collected from your group and from other classmates.

When you are finished, check with your instructor, then return mobile animals to an undisturbed portion of the fouling community growing on the dock. However, most sessile animals will not be able to reattach, themselves, so they will be returned to the ocean to serve as food for other organisms. In harbors and marinas, docks and boat bottoms are scraped periodically and it is likely that the organisms you examined would have met the same fate anyway at a later date.

FOULING COMMUNITY DATASHEET

Name(s): ______________________________

Location: ____

Date, time: ______________________________

Water temperature: _____°C, Salinity: ________

Phylum	*Common name(s)*	*Number in your sample*	*Class total for this phylum:*	*% of total*	*Feeding method*
Porifera					
Platyhelminthes					
Annelida ** any worm with long bristles should be picked up with a spoon.*					
Arthropoda					
Mollusca					
Echinodermata					
Cnidaria					
Chordata					
Bryozoa					
		Grand total of all organisms:		100%	

Using a spreadsheet program, create a pie chart representing the proportions of major taxa found in each of the different locations where you sampled. Prepare a report comparing community composition on the shady vs. sunny side of the dock, and/or a shallow vs. deeper community.

In your report, include the following elements:

1. Identify the dominant phyla in your samples. Which phylum contained the most individuals of one species? Which phylum contained the most species?
2. Discuss whether your samples represent a mature fouling community or a younger community in an earlier stage of ecological succession. Give 3 observations about your sample to explain your answer.
3. Make a diagram of a food web using specimens from one of the communities you sampled, indicating which are scavengers, predators, filter feeders, and herbivores.
4. Offer hypotheses to explain any observed differences in species diversity or abundance between the different locations that were sampled.

WEB EXPLORATIONS FOR FURTHER STUDY:

Explore relevant websites to stimulate interest in current fouling research.

Office of Naval Research http://www.onr.navy.mil/

Conducts marine research to support science and technology programs of the U.S. Navy and Marine Corps.

U.S. Geological Survey Coastal Marine Geology Science Centers:

Woods Hole Science Center http://woodshole.er.usgs.gov/index.html

St. Petersburg/Center for Coastal & Watershed Studies http://coastal.er.usgs.gov/

Western Coastal and Marine Geology http://walrus.wr.usgs.gov/

The U.S. Geological Survey is our nation's largest agency that undertakes large-scale investigations in biological and natural science, providing information about our water, energy, and mineral resources.

Activity: Search the ONR site and the USGS site closest to your state to research the answers to these questions:

1. How are fouling communities a problem for boat owners and for international shipping?
2. How much of an economic impact do fouling organisms have on maritime commerce and/or the Navy in the United States?
3. What are some technologies currently being researched to reduce the effects of marine fouling on vessels and undersea structures?
4. Research the spread of non-native tunicate species along the East and West Coasts. Two important invasive types are *Didemnum* and *Botrylloides*. How did these sessile animals arrive in the areas they now colonize? Why are they considered serious threats to native marine fauna?

ILLUSTRATED GUIDE TO FOULING COMMUNITY ANIMALS

Use this guide to help identify the phyla and classes to which your animals belong. Use regional guides specific to your area for more information on local species.

1. Animal is **sessile**, growing like an encrusting sheet over the substratum, or in irregular clumps:

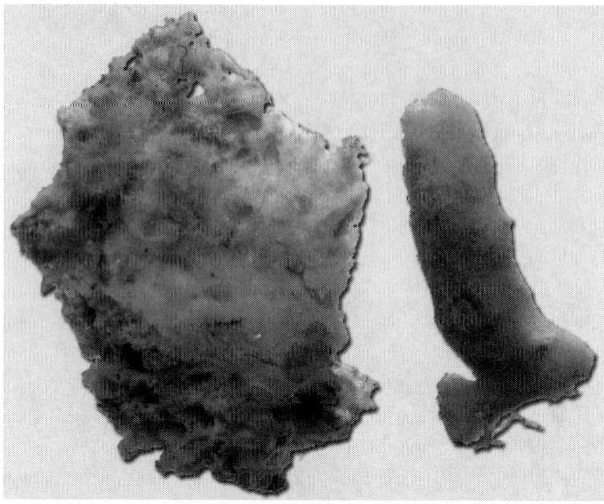

Figure 2. Sponges

Phylum Porifera: Sponges
Spongy texture, with numerous pores present throughout the body.

DVD: Color pictures of these organisms are provided in the DVD chapter for this field study.

Phylum Chordata, Class Ascidia: Colonial Tunicates The colony is composed of numerous sac-like **zooids;** each zooid is a separate clonal animal, but zooids are connected to each other by a tissue connection at their bases.

Figure 3. Colonial tunicates. A: Colony with intergrowing sponges. Arrow points to zooids, magnified in b. C: Zooids filter-feeding, with open incurrent and excurrent apertures.

Phylum Chordata, Class Ascidia: Compound Tunicates

Individual zooids are encased in a sheet of gelatinous, rubbery tissue that covers the whole colony; zooids usually are arranged in circles around a larger hole in the rubbery covering.

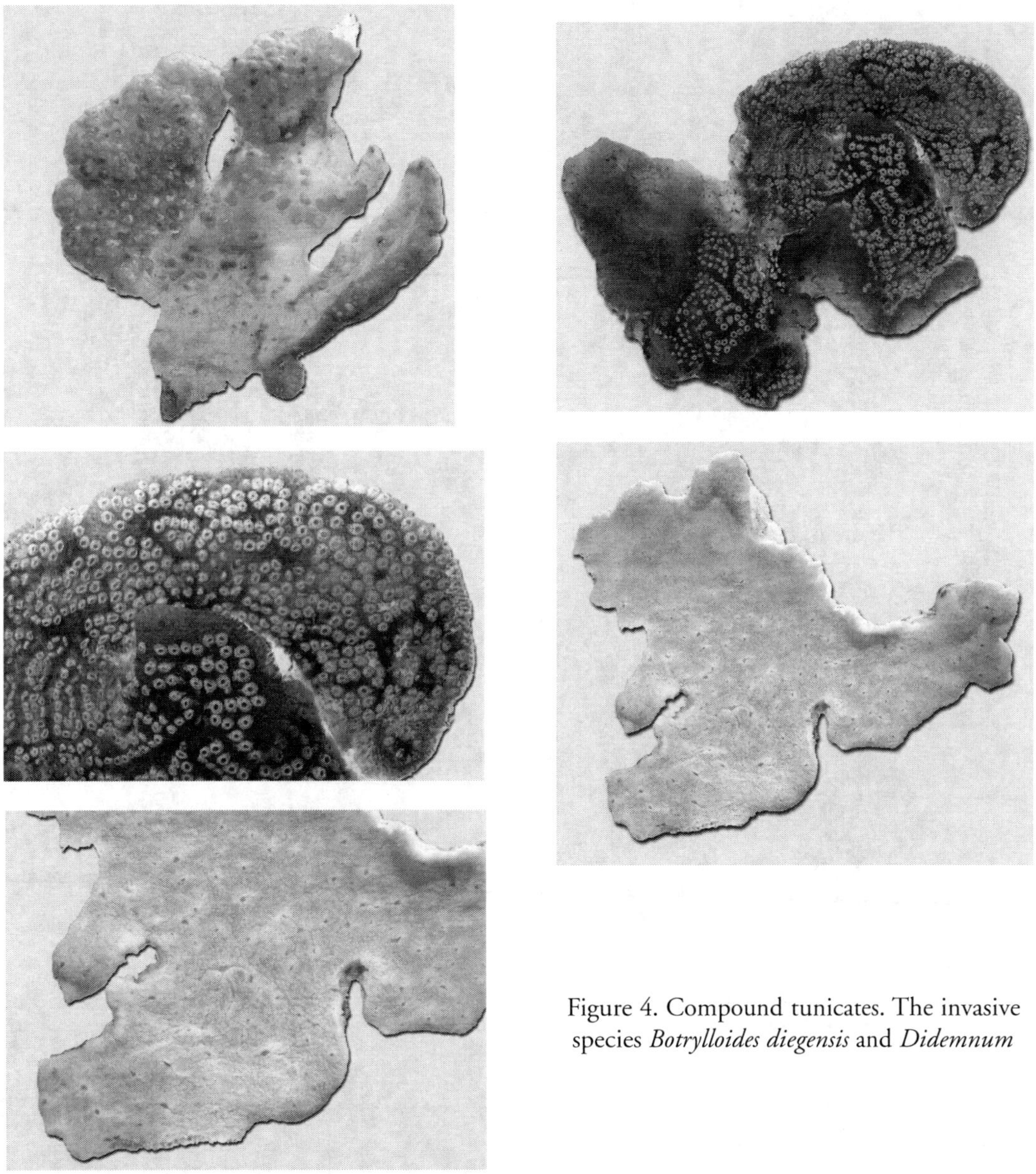

Figure 4. Compound tunicates. The invasive species *Botrylloides diegensis* and *Didemnum*

Phylum Ectoprocta: Encrusting Bryozoans (moss animals)

Individual zooids are housed in regular, box-shaped containers that feel crunchy; they grow in a lacy sheet over surfaces.

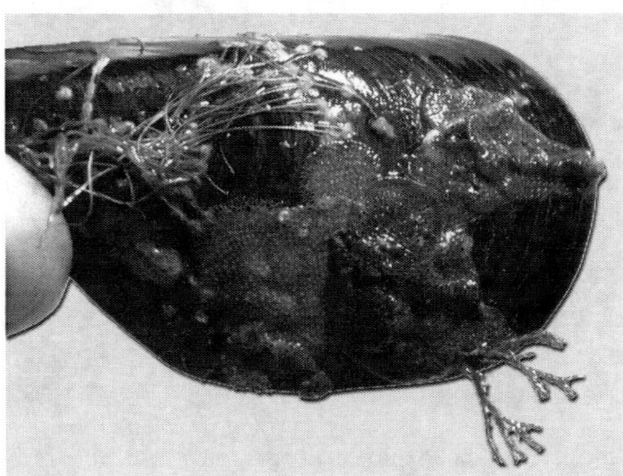

Figure 5. Encrusting bryozoans on a mussel shell.

2. Animal colony looks like a bushy plant, or one with fine filaments branching out from a stalk, and can be various colors:

Phylum Cnidaria, Class Hydrozoa: <u>Hydroids</u>

Tiny **<u>polyps</u>** may be visible on the stalks, each polyp bearing numerous tentacles; colony may appear feathery or filamentous.

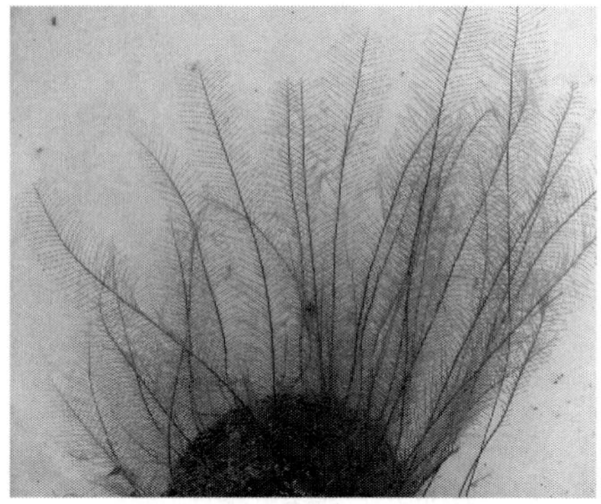

Figure 6. Hydroid colonies.

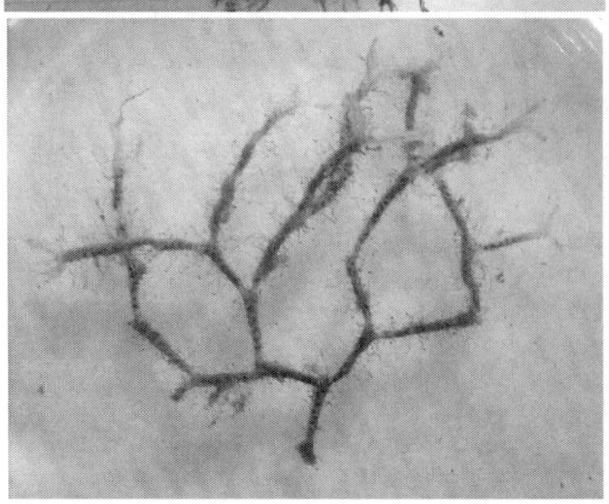

Phylum Ectoprocta: Arborescent <u>Bryozoans</u> (moss animals) (Latin: *arbor*, a tree)

Look closely and see the individual zooids growing together in the "branches."

Figure 7. Bryozoans. These may resemble hydroids, but bryozoans do not string if touched.

3. Animal is a sessile individual, living on a stalk or within a tube:

Phylum Annelida, Class Polychaeta: Tubeworms

Tube is **calcareous** or made of organic **detritus,** such as bits of seaweed; a feathery fan may extend from one end of the tube, but retracts when disturbed; or animal has long, spaghetti-like tentacles extending in all directions.

Figure 8. Tubeworms. Top left: A calcareous fan worm and a feather duster worm (Body length: 15 mm); Above: A coiled Medusa-head worm removed from its tube, showing feeding tentacles (Size: 3 cm).

Phylum Cnidaria, Class Anthozoa: Sea anemones

Mouth is within the center of a ring of smooth tentacles; tentacles retract when disturbed and are sticky to the touch.

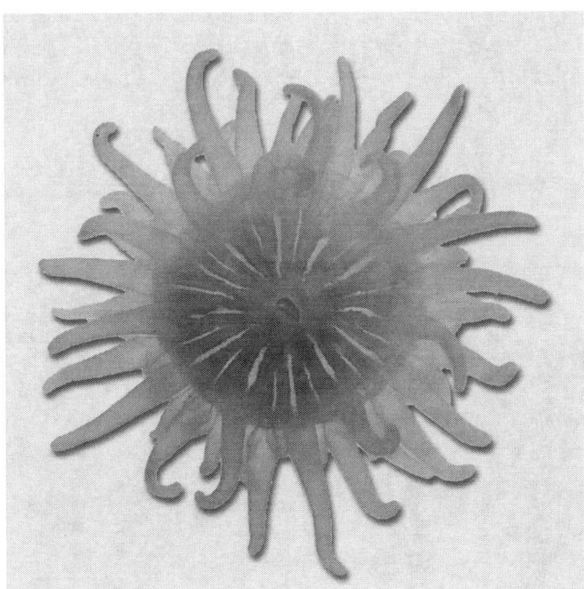

Figure 9. The Brooding sea anemone, *Epiactis prolifera.* Size: 1 cm.

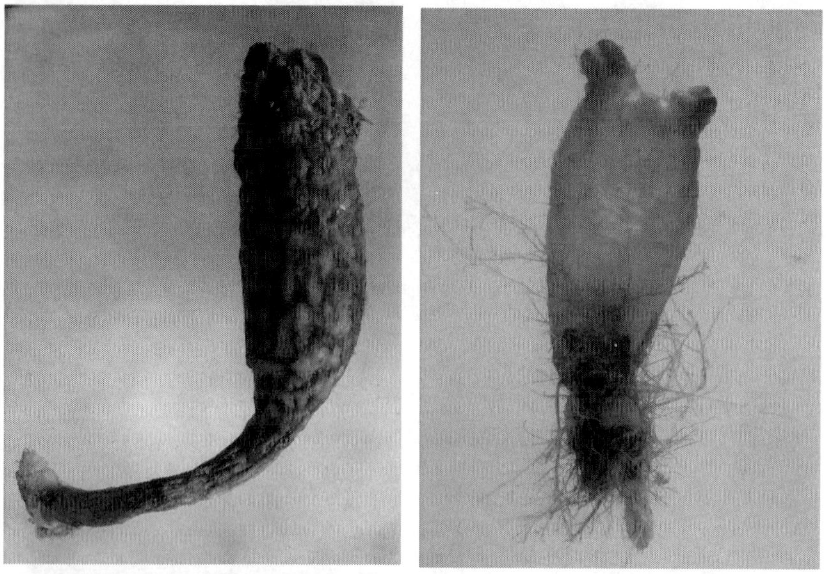

Phylum Chordata, Class Ascidia: Solitary Tunicates (sea squirts)

Animal is housed within a rubbery or leathery covering (a "tunic") that has two bumps at the top; when the tunic is gently squeezed, water squirts out of the bumps.

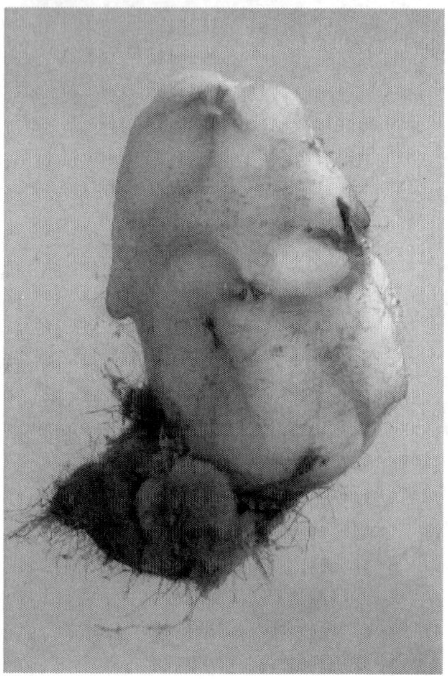

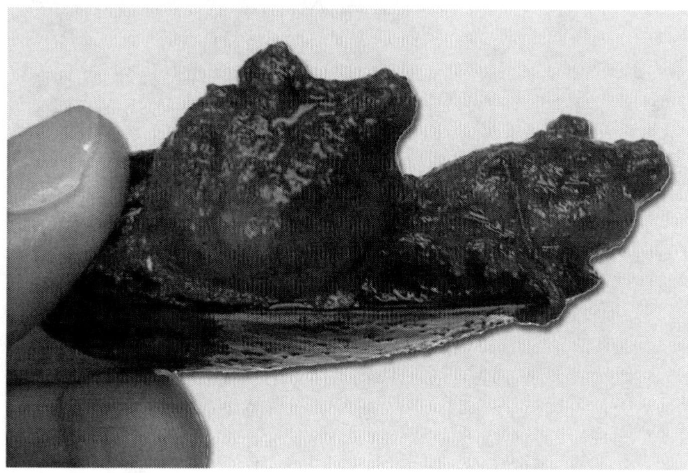

Figure 10. Solitary tunicates. Top left: Stalked tunicate, *Styela clava* (Size: 12 cm). Top right: Orange Sea Peach, *Halocynthia aurantia* (Size: 6 cm). Bottom left: *Styela plicata* (Size: 3 cm).

4. The animal resembles a sheet of crawling tissue; the body is flattened dorsoventrally ("top to bottom") and is <u>very thin</u> (<1 mm thick,) and is **<u>not</u>** round in cross-section; body lacks segments:

Figure 11. Flatworm (Size: 5 mm).

Phylum Platyhelminthes (marine flatworms)

5. The body is elongated, wormlike AND has segmented regions; on each segment there are flap-like structures (**"parapodia"**) that may have tufts of bristles sticking out:

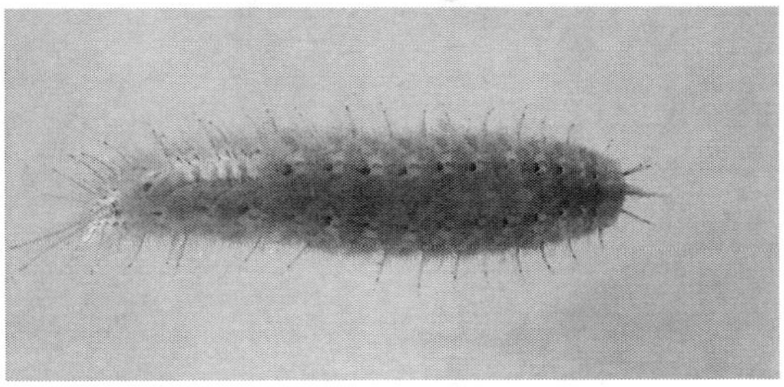

Figure 12. A scaleworm, a marine annelid (Size: 20 mm).

Phylum Annelida, Class Polychaeta (marine annelid worms)

6. The organism is a **bivalve** (it has a hard shell consisting of two halves, or **valves**) **Phylum Mollusca, Class Bivalvia:**

Figure 13. Mussels

Shells are dark-colored, valves are symmetrical in size, and have brown hair-like threads (**byssal threads**) attached to them at one end of the shell (Mussels).

Figure 14. Oysters

Shells are generally lighter in color; valves are irregular in shape; the lower valve is cemented to the substratum (Oysters).

7. The organism is mobile, with a hard shell (exoskeleton) that is segmented AND there are paired, jointed appendages (such as legs and antennae): **Phylum Arthropoda, Class Crustacea**

Juvenile crab

Figure 15. Juvenile shore crab, *Pachygrapsus crassipes.*

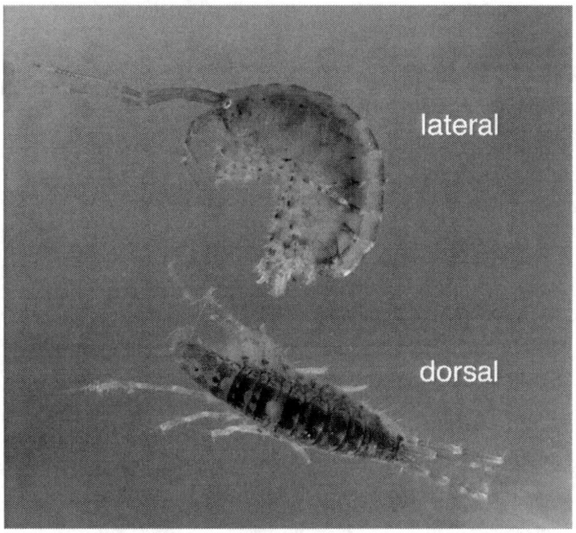

Body is compressed **laterally** (animal is taller than it is wide): **Amphipods**

Figure 16. Amphipods. Size: 5 mm.

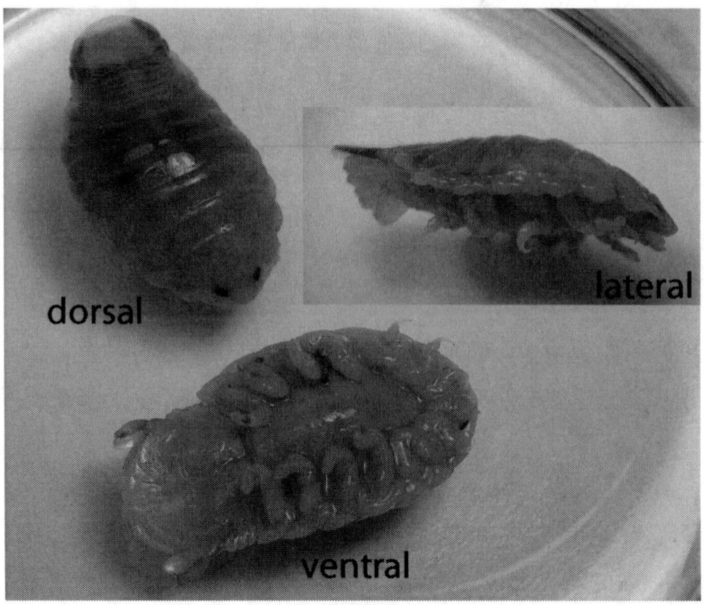

Body is flattened **dorsoventrally** (animal is wider than is tall, and resembles a pillbug or sowbug): **Isopods**

Figure 17. Isopods. Size: 1 cm.

8. Animal has 5 arms (some may be broken off) radiating out from a central flat disk:

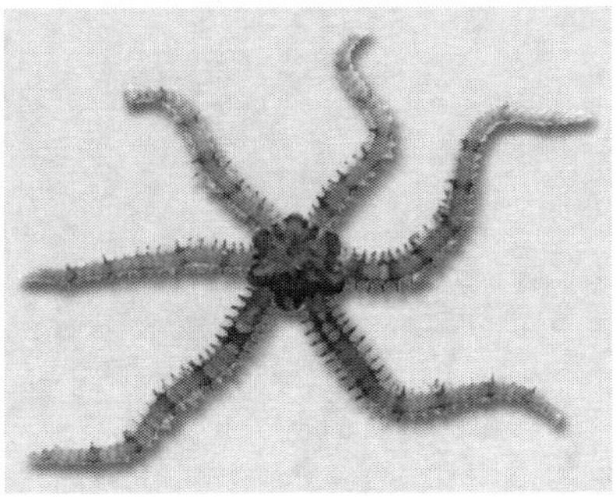

© P. Detwiler 2008

Figure 18. Brittle star

Phylum Echinodermata, Class Ophiuroidea: Brittle Stars

FIELD STUDY 3

Rocky Intertidal Zone

INTRODUCTION:

The rocky intertidal zone is a coastal habitat that never ceases to fascinate and delight those who carefully observe its secrets. Many marine biologists admit that their sense of wonder and curiosity about life in the ocean was first sparked by a visit to the seashore and a glimpse into the underwater world within a tidepool. How will a visit to the rocky intertidal zone affect you?

The rocky intertidal zone has been studied extensively by marine biologists throughout the twentieth century, and the results of numerous experiments have yielded much ecological knowledge about the processes and interactions affecting the abundance and distribution of marine plants and animals in the rocky shore community. Yet many questions remain unanswered. During this field study period, you will survey this intertidal region as an ecologist would: You will observe species in detail, looking closely and carefully at anatomical structures and hypothesize their adaptive value to the organism. Also, you will conduct surveys through the different intertidal levels to become familiar with the habitat, collect data on common species, create graphs of your results, and interpret them using ecological thinking.

The goals of this lab are for you to accomplish the following:

- Identify the common seaweeds and animals of the rocky intertidal zone.
- Observe the zonation patterns of characteristic species inhabiting your local area.
- Conduct a quantitative survey of intertidal zone organisms using quadrat sampling.
- Describe the processes that lead to the formation of horizontal bands in the intertidal zone.
- Explain how different organisms are adapted to the particular environmental conditions of the intertidal zone.
- Explain several ways to reduce your impact when visiting this habitat.

BACKGROUND:

The intertidal zone is the region of coastline covered by the highest tides and exposed by the lowest tides. Seashore life is strongly affected by the rise and fall of the tides. Once or twice a day (depending on whether the tide cycle is **diurnal** or **semidiurnal**), the incoming tide covers the shoreline, bringing oxygen, food, and dissolved nutrients to the organisms that live there. When the tide recedes, many of the organisms are left exposed to the elements until the next high tide comes in. This alternating pattern of submergence and exposure causes organisms to be distributed at certain tidal heights throughout the intertidal zone, on the basis of how well they tolerate exposure to air and changes in other **abiotic** factors that vary between high and low tide.

Most organisms of the rocky intertidal zone live as benthic epifauna on rocks, within crevices, or beneath algae. Species diversity (the number of different species living within an area) is also influenced by tidal height. The rocky coastline is separated into 4 subzones, each with a characteristic suite of organisms:

Zone 1: The **Splash** or **Spray Zone** (also called the **Supralittoral Zone**) is the uppermost zone and receives moisture as spray during the highest of tides and as splash during storms. Rocks are typically covered with dark patches of cyanobacteria (blue-green algae).

Zone 2: The **High Zone** may be covered only once a month during the highest tides. Organisms in this region have evolved adaptations to deal with the problems of **desiccation**. Acorn barnacles and tiny snails called periwinkles are common.

Zone 3: The **Middle Zone** is regularly exposed and submerged during each tidal cycle. More species are found here, and open space (areas of bare, solid rock) is in short supply, so competition among **sessile** organisms for places to attach and grow is strong. Because of this, many smaller sessile organisms such as barnacles and bryozoans are seen growing upon the hard surfaces of larger animals like mussels and crabs.

Zone 4: The **Low Zone** is the region exposed only during the lowest of low tides, when tide tables show a "minus" tide. Organisms in this zone are the least tolerant of exposure, since they lack anatomical structures to prevent water loss. Sometimes subtidal animals venture into the low and middle zones during high tide and get stranded within tidepools when the sea retreats; they must then wait for the next high tide to flood the pool, allowing them to return to sea.

PROCEDURE:

First and foremost are safety concerns: Wear footwear with good traction that covers your feet (no flip-flops), and watch where you step! Deep tidepools may be covered by floating algae and/or seagrasses, so do not step anyplace where you cannot see the bottom; you may step into a deep hole and twist your ankle. Algae-covered rocks are slippery, so always try to step on bare rock and make sure your footing is firm before you take another step. Do not venture into the surf zone, and never turn your back to the ocean — even on a calm day. Rogue waves (also called sneaker waves) are larger than the usual sets and can suddenly crash onshore, knocking you down.

Look for organisms within tidepools, in crevices, under clumps of seaweed, and around the margins of large rocks. Many intertidal organisms are small and camouflaged, and none will advertise their presence to you, so you need to lower yourself down to the water level and look closely to discover these small creatures. It is best to observe the natural behaviors of tidepool animals when they are undisturbed. Do not remove **sessile** organisms (animals and algae that are attached to the rocks). If allowed by the regulations governing your study location, you may pick up mobile crustaceans and place them in a container of seawater for closer observation. Soft-bodied animals (snails and sea slugs) and echinoderms (sea urchins and sea stars) attach themselves to rocks with delicate tissues, and these should not be picked up. Be very gentle when handling any marine animal to avoid unintentionally damaging it. Some intertidal animals have very specific diets and are only found in specific habitats where their prey is located. If you pick one up and return it to a different location, you may separate it from its food source, which it might not find again, or you may place it in an unsuitable microhabitat where it could quickly die. Because we want to minimize our impact on the rocky shore community:

Each animal that you pick up for study must be returned to the **exact spot where you found it.**

Do not move or turn over rocks; the animals living underneath **will be crushed,** and if a rock is turned over but is not returned to its original position and orientation, sessile animals attached to the underside will die from the drying effects of sun and air.

Keep track of the time so you can finish all the activities, and do not lose any equipment on the rocks. At the conclusion of the field trip, the class will reconvene to share data and discuss results.

Date: ____________________

Time: ____________________

Time of the Low Tide Today: ____________________

Beaufort Sea State (from Appendix E): ___________

Air Temperature: ______________°C

Ocean Temperature: ______________°C

Seawater Salinity (ppt): ______________

Location (GPS coordinates or detailed description):

General Weather Conditions:

ACTIVITY 1: SURVEY OF INTERTIDAL ZONE AND SCAVENGER HUNT

You and your partner will investigate the different zones of the intertidal area; making it a goal to see ***with your own eyes*** at least 20 species of marine algae and animals. Check off the species as you encounter them during your survey of the intertidal zone.

CHECKLISTS OF ROCKY INTERTIDAL ORGANISMS (Common genera are provided in italics)	
PACIFIC COAST:	ATLANTIC COAST:
q Abalone (*Haliotis*)	q Acorn barnacles (*Balanus*, *Chthamalus*)
q Acorn barnacles (*Balanus* and *Chthamalus*)	q Bladder wrack (*Fucus*)
q Black turban snail (*Tegula*)	q Blue mussel (*Mytilus*)
q Brittle Star	q Chiton (*Ischnochiton*)
q Broken-backed shrimp	q Coralline algae (both erect and crustose kinds)
q California spiny lobster (*Panulirus*)	q Dogwhelk (*Nucella*)
q Chitons	q Dulse (*Rhodymenia*)
q Coast mussel (*Mytilus*)	q Frilled anemone (*Metridium*)
q Coralline algae (both erect and crustose kinds)	q Green sea urchin (*Strongylocentrotus*)
q Elegant (clonal) anemone (*Anthopluera*)	q Green shore crab (*Carcinus*)
q Feather boa kelp (*Egregia*)	q Hermit crab (*Pagurus*)
q Giant Kelp (*Macrocystis*)	q Irish Moss (*Chondrus*)
q Gooseneck barnacle (*Pollicipes*)	q Knotted Wrack (*Ascophyllum*)
q Green anemone (*Anthopleura*)	q Limpets (*Acmaea*)
q Hermit crab (*Pagurus*)	q Periwinkle (*Littorina*)
q Kelp crab (*Pugettia*)	q Rock crab (*Cancer*)
q Limpets (*Collisella*, *Acmaea*)	q Scud (*Gammarus*)
q Nudibranchs (sea slugs)	q Sea lettuce (*Ulva*)
q Octopus	q Sea star (*Asterias*)
q Periwinkle (*Littorina*)	q Slipper shell (*Crepidula*)
q Purple and Red sea urchins (*Strongylocentrotus*)	q Sponge weed (Dead man's fingers, *Codium*)
q Purple shore crab (*Pachygrapsus*)	q Sponges
q Rockweed (*Pelvetia*)	q Tidepool shrimp (*Praunus*)
q Sea Hare (*Aplysia*)	
q Sea lettuce (*Ulva*)	
q Sea Palms (*Postelsia* and *Eisenia*)	
q Sponge weed (Dead man's fingers, *Codium*)	
q Sponges	
q Surfgrass (*Phyllospadix*)	
q Tidepool sculpins	

Additionally, your instructor will assign you and your partner one particular organism to carefully observe and study. Your instructor may require you to turn in your field notes about this organism. Record your Field Notes below.

Scavenger Hunt Organism Name:

What Phylum and Class does it belong in?

Color/Texture: Size (mm):

On what type of **substrate** is it found—is it **epifaunal**, living on rocks, algae, or on another animal? Is it **sessile** or mobile?

In what intertidal zone is it found (1—2—3—4)? Is it a colonial or a single organism?

Does it appear aggressive or passive? Is it aware of your presence? Does it try to escape from you?

Does it have any unusual body structures? Indicate any in your drawing below.

Is it showing any particular behavior that might give you a clue toward how it feeds and what it eats?

Draw your organism in the box below, noting details of its anatomy.

Other notes/observations:

ACTIVITY 2: MOLLUSK MEASUREMENTS

Your instructor will show you two locations where a common species of marine mollusk is abundant. At both locations, measure to the nearest mm the shell length of all the individuals of that species found within a 1-m diameter circle. Record the sizes of <u>all</u> individuals of that species, not just the largest ones! While doing this, learn to distinguish 2 other species of mollusks found in the same environment by making detailed drawings of those species with notes on color, size, shell features, etc. in the boxes below. When you have measured at least 15 specimens, give the meter stick and plastic ruler to another student who is ready to do this activity. Back in the laboratory, this data will be analyzed as your instructor advises.

Name of mollusk species: ______________________________

Site #1 Description of study location:

Shell lengths (in mm):

Site #2 Location description:

Shell lengths (in mm):

Are any of these mollusks consistently found in a particular type of **<u>microhabitat,</u>** such as crevices, depressions in rock, under algae? If so, describe the microhabitat(s) that they seem to prefer.

Are the animals found singly, or are they clustered together in groups?

Species 1:	Species 2:

ACTIVITY 3: COMPARISON OF TIDEPOOLS: ABIOTIC AND BIOTIC FACTORS

Select two tidepools to investigate, one in the upper intertidal and the other in the lower zone. Estimate the difference in vertical height separating the two tidepools. Using the measuring tools provided by your instructor, distinguish at least 5 differences between the tidepools.

	Tidepool A	Tidepool B
	Found in Zone #_______	Found in Zone #_______
Water temperature, °C		
Salinity, ppt		
Dissolved oxygen (DO), ml/l		
pH		
Depth, cm		
Amount of exposure to wave shock		
Dominant algal species		
Dominant invertebrate species		
Diversity of species (high–low)		
Other		

Which pool more closely resembles the ocean? Was this pool more diverse in species than the other pool?

Are there noticeable differences in the composition of tidepool species (for example, does one tide pool contain just coralline algae and a few snails and the other pool contain many different types of species)? Propose at least 2 hypotheses that may help explain any differences:

ACTIVITY 4: QUADRAT SURVEY ALONG RANGE OF INTERTIDAL HABITAT

A **transect** is a type of study where data are recorded along a continuous path. For this activity, the transect will extend from the low zone to the high zone, following the contours of the substrate. Starting at the beginning of the transect, gaze along its length up to its ending point in the high zone. Can you visually detect any boundaries between the zones? In many cases, those boundaries are sharply defined by conspicuous indicator organisms, but in some cases, the distinctions between zones are less obvious.

The purpose of this activity is to record the patterns of abundance and distribution of the dominant organisms across the intertidal zone.

PROCEDURE:

1. Your instructor will tell you the names of 10 target species to record in this exercise. Enter them in the column marked "Species" in the Datasheet on the following page.
2. Place a quadrat along the beginning of the transect line (at meter mark 0), and identify and count the number of organisms of each target species found within the quadrat. If organisms are too numerous to count within the entire quadrat, just count those within a quarter of the quadrat and multiply by 4. Record your data in the Datasheet.
3. Move the quadrat to the next meter mark and repeat, working your way up the transect line toward the spray zone. Check with your instructor to see how far to continue along the transect line.

DATA ANALYSIS AND GRAPHING:

1. To standardize the data, for each of the target species, convert the number of individuals per quadrat into density per square meter (m^2).
2. On the Results sheet (Figure 1) draw a horizontal line along the middle of each species' row, showing the range of where each was observed along the transect line. Figure 1 shows an example species that was present along the transect between quadrats 0–9, but was absent at meter 10 and beyond.
3. For species 1-10 on Figure 1, the height of a species' row will represent that species' maximum density per m^2 observed along the transect. Draw a vertical line at each meter mark with its height proportional to the density of that species recorded at that meter mark. In Figure 1, the maximum observed density of the example species seen along the entire (length of the) transect was 100 individuals per m^2. At meter mark 4, the arrowed line extends the entire height of the row, indicating that the maximum density was observed there. Similarly, at meter mark 1, the arrowed line covers one-quarter of the row height, corresponding to a density of 25 individuals per m^2 at that recorded location.
4. Draw lines connecting the vertical lines together to make a "kite diagram," as shown in Figure 1.

Interpret your completed diagram to answer Question 6 on the Review Sheet.

Datasheet for **Activity 4:** Quadrat Survey Along Range Of Intertidal Habitat

Species:	Quadrat (Number of Meter Mark on the Transect Line)																				
	0	1	2	3	4	5	6	7	8	9	10	11	12	13	14	15	16	17	18	19	20
1. Example species:	10	25	75	90	100	80	60	30	20	10	1	0	0	0	0	0	0	0	0	0	0
2.																					
3.																					
4.																					
5.																					
6.																					
7.																					
8.																					
9.																					
10.																					
11.																					
12.																					
13.																					
14.																					
15.																					
16.																					
17.																					
18.																					
19.																					
20																					

Seaweeds must be attached to the substrate within a quadrat to be counted; do not record detached fragments in a quadrat.
If you cannot identify the organism in the field using a regional guide for your area, take a digital photo for later identification.

Figure 1. Results Sheet for Activity 4. Vertical Zonation Patterns of 10 Dominant Intertidal Species.

Low Zone — Range across Intertidal Zone (Meter Mark of Transect Line) — High Zone

	0	1	2	3	4	5	6	7	8	9	10	11	12	13	14	15	16	17	18	19
Example: 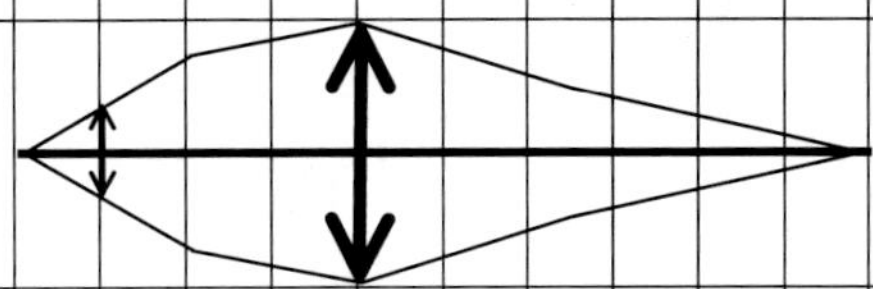																				
1.																				
2.																				
3.																				
4.																				
5.																				
6.																				
7.																				
8.																				
9.																				
10.																				

Density (per m^2)

CLASS SUMMARY: SCAVENGER HUNT RESULTS (FROM ACTIVITY 1)

Tidal zone (circle one): 1 = Splash Zone; 2 = High; 3 = Mid; 4 = Low

Common name:	Main Taxa: (Phylum, Class)	Tidal zone (1 2 3 4) Microhabitat	Trophic level: What it eats, and how it gets its food
1. *Example:* Limpet	Mollusca: Gastropoda	(1) 2 3 4 On rocks and in crevices	Herbivore Scrapes algal film off of rocks with radula
2.		1 2 3 4	
3.		1 2 3 4	
4.		1 2 3 4	
5.		1 2 3 4	
6.		1 2 3 4	
7.		1 2 3 4	
8.		1 2 3 4	
9.		1 2 3 4	
10.		1 2 3 4	

Common name:	Main Taxa: (Phylum, Class)	Tidal zone (1 2 3 4) Microhabitat	Trophic level: What it eats, and how it gets its food
11.		1 2 3 4	
12.		1 2 3 4	
13.		1 2 3 4	
14.		1 2 3 4	
15.		1 2 3 4	
16.		1 2 3 4	
17.		1 2 3 4	
18.		1 2 3 4	
19.		1 2 3 4	
20.		1 2 3 4	

Use the information from this summary table to complete Question 5 on the Review Sheet.

Name: ______________________

Date: ______________________

Section: ______________________

REVIEW:

1. Hypothesize why the mollusks you studied were found in particular microhabitats. Is there a survival benefit the mollusks may gain by congregating together?

2. Refer to your answers to Activity 3. Other than temperature, what other physical variables may fluctuate in a tide pool over the course of a low tide cycle, and how can these changes affect tide pool organisms (both their abundance and the diversity within a tide pool)? Discuss at least 3 variables.

3. How do you think the intensity of physical/chemical factors is related to species diversity? (To answer this, would you expect many different species to exist in an environment with harsh physical/chemical factors?) Which tidal zone is <u>least</u> diverse, and why?

4. How would the distribution and abundance of certain organisms be affected if the weather conditions were very different from what you experienced today (i.e., think of what types of animals would be most affected, and in what ways, if it were a very sunny, hot day versus a cold, rainy, overcast day)?

5. Construct a food web in the table below for at least 10 of the organisms from the Scavenger Hunt Activity. Place the name of each organism at its proper trophic level (i.e., identify each as either a **producer**, **primary consumer**, **secondary consumer**, or **tertiary consumer**). Draw arrow(s) connecting the organisms in one trophic level to their predators on the next level to indicate the direction of energy flow through the rocky intertidal community.

Trophic Level	Category of Organisms at this Level:	Examples of Species:
4	Tertiary Consumers (Carnivores):	
3	Secondary Consumers (Carnivores):	
2	Primary Consumers (Herbivores):	
1	Producers:	

6. What general patterns are revealed in your kite diagram? Interpret your diagram and identify any organisms that are widely distributed across tidal zones (consider both animals and seaweeds). On a separate piece of paper, discuss some possible reasons why other species have a more narrow distribution. Can you think of any particular physical factors that make the widely distributed organisms more tolerant of variations in the environment than the narrowly distributed organisms? Think about the importance of biological factors such as competition and predation. Would the distribution of any species in your kite diagram be affected by the presence of any other intertidal species? If so, which would be affected, and in what ways?

7. Summarize the adaptations of intertidal organisms to environmental stresses by filling in the blanks in the following table.

Stress:	Organism Examples:	Adaptations to this Factor:
Exposure to air and desiccation	1. Barnacles	1.
	2. Anemone	2.
	3. Limpet	3.
	4. Crabs/shrimp	4.
Strong waves crashing upon rocks and tidal surge	1. Kelp	1.
	2. Coralline algae	2.
	3. Sea stars	3.
	4. Mussels	4.
	5. Barnacles	5.
Intense competition for attachment sites between sessile organisms	1.	Fast growth rate
	2.	Overgrowth of other forms
	3.	Aggression/fighting to claim space

APPENDIX A

Metric System

The United States has retained an archaic system of measurement that is an anomaly in the twenty-first century. All nations—except the USA, Liberia, and Myanmar (formerly called Burma)—have adopted the metric system, a common language for science, commerce, and day-to-day life for billions of people around the world. Because of its worldwide usage, the metric system is called the **International System of Units**, abbreviated **SI**, and all scientific data are expressed in SI units. The metric system is easy to learn because it is based on a decimal system (based on units of 10).

Which system of measurement is easier to work with? The answer is obvious when you look at the two columns of distance measurements below. Which column is easier and faster to solve for total length? *Note*: The answer to both columns is the same.

English System Units:	**Equivalent Metric System Units:**
1 yard, 11-5/8 inches	1.209 meters
2 feet, 7 inches	0.787 meters
2 yards, 4 feet, 3 inches	3.124 meters
1/2 mile, 414 feet, 9-3/4 inches	931.106 meters
______________________	____________

Total length in feet: ____________ Total length in meters: ____________
Total length in yards: ____________
Total length in miles: ____________

Despite its non-intuitive and illogical system of units, English weights and measures are firmly entrenched in most Americans' minds; however, with practice, the metric system can be learned easily and quickly. To gain a mental sense of the units, learn to associate everyday objects with their metric measurements, and use common objects as "rulers" to visualize the size of SI units.

In the metric system, the basic unit of length is the meter, volume is the liter, mass is the gram, and temperature is degrees Celsius. Examine the chart below, which presents the scale of each prefix. Of the various prefixes listed, three must be learned thoroughly (k, c, and m) because the dimensions and weights of most objects commonly encountered on a daily basis occur within those size range.

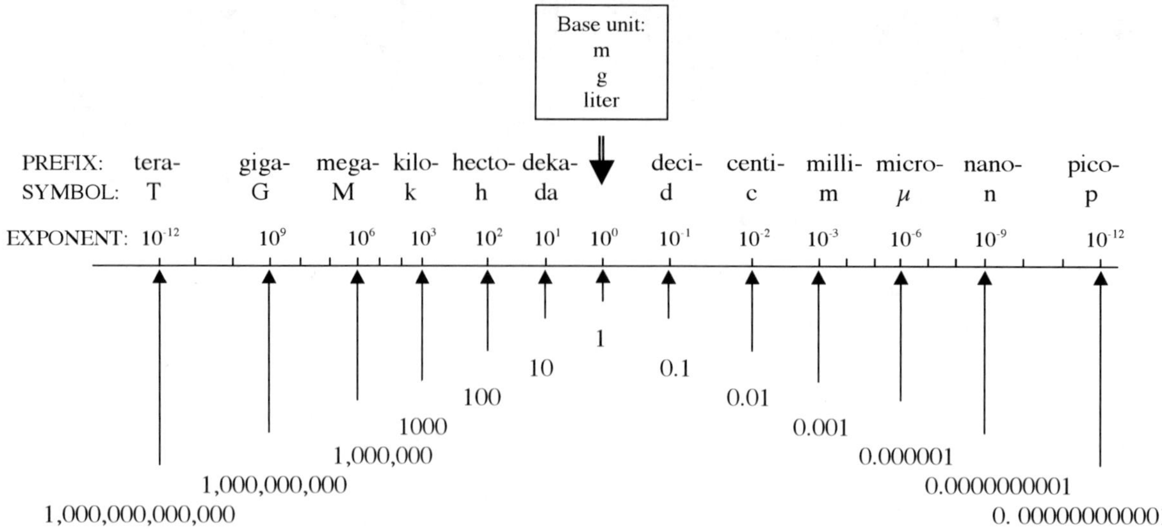

Converting metric units is simple. One easy method to convert units within the metric system is the **Step System**.

1. First, determine the number of steps (exponents) between the unit that you wish to convert *from*, to the unit you wish to convert *to*. For example, there are 3 steps between a kilometer (10^3) and a meter (10^0).

2. If you wish to convert a **larger** unit to a **smaller** unit, move the decimal point of the larger measurement the appropriate number of steps to the *right*.

***Example 1*:** The prefix *k* refers to a factor of 1,000, which is a larger number than the desired unit, the meter. Therefore, move the decimal point 3 places (or steps) to the right to arrive at the desired unit: 1.0 km = 1,000.0 m

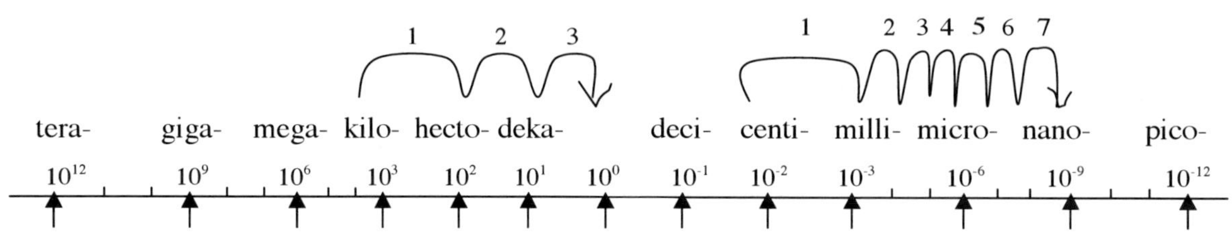

Example 2: Convert 1 cm to nanometers. Note that there are 7 steps (or exponents) between centimeters (10^{-2}) and nanometers (10^{-9}). Move the decimal point 7 steps to the right: 1 cm = 10,000,000 nm.

3. To convert from a **smaller** unit to a **larger** unit, move the decimal point of the smaller measurement the appropriate number of steps to the *left*.

***Example 3*:** Convert from centimeters to meters. First, recognize that the prefix *c* refers to a factor of 0.01, which is a smaller unit than the desired unit you are converting to (the meter); secondly, notice the units are separated by 2 steps (exponents). Then, move the decimal point 2 places (or steps) to the left; 1.0 cm = 0.01 m

To remember which direction to move the decimal point, visualize this image:

Larger
Left

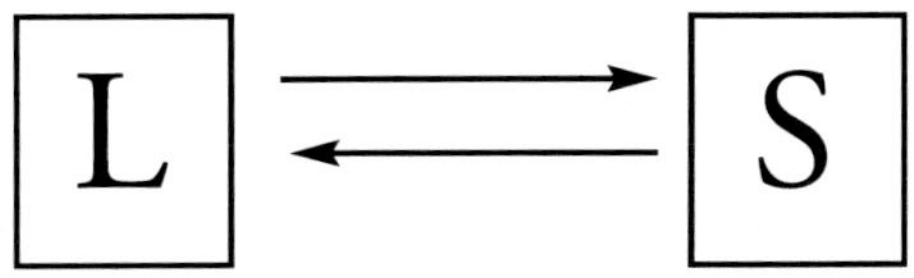

Smaller
Starboard (which is the right side of a ship)

Convert the following distances from the given initial unit of measurement into the other.

Larger Unit to Smaller Unit:				Smaller Unit to Larger Unit:			
1 cm	=	____________	mm	1 μm	=	____________	mm
1 m	=	____________	mm	1 nm	=	____________	μm
1 km	=	____________	m	1 cm	=	____________	km
1 g	=	____________	mg	1 μg	=	____________	g
1 kg	=	____________	g	1 μg	=	____________	mg
1 l	=	____________	ml	1 ml	=	____________	l

Now that you are familiar with the step system, convert these:

29 μm	=	____________	cm
31.6 cm	=	____________	μm
9 μm	=	____________	mm
165.345 m	=	____________	km
56.21 g	=	____________	kg
91.4 ml	=	____________	l

VOLUME:

Volume can be expressed in two ways, as a cubic function of distance (length x width x height) and by using the basic unit for volume, the liter (L or l). The milliliter is defined as the volume of water contained within a cube, with each side measuring 1 cm in length.

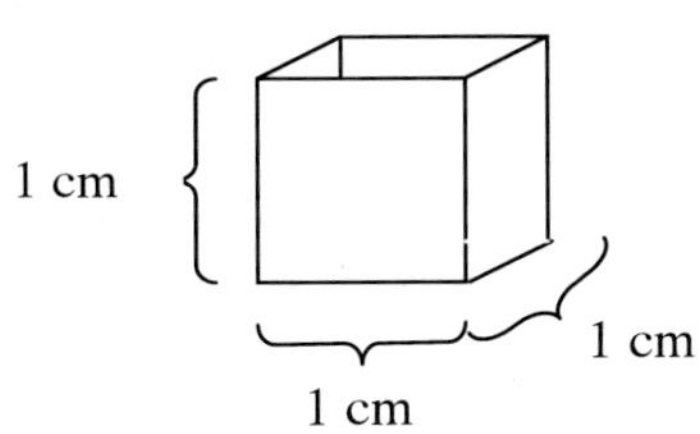

1 cm x 1 cm x 1 cm = 1 cm^3 = 1 ml

How many ml are in 1 L?

Therefore, how many 1 cm^3 cubes can be fit into a larger cube having a volume of 1 L^3?

MASS:

Mass, or weight, is expressed as multiples of the **gram** (g). The gram was originally defined as the weight of 1 cm^3 (= 1 ml) of pure water at a temperature of 4° Celsius. At this temperature, water is at its maximum density. Given this information, answer the following questions.

How much should 1 liter of water at 4°C weigh in g? ____________
How many grams does 0.345 l of water at 4°C weigh? ____________
How many milligrams would 1.32 l of 4°C water weigh? ____________
How many milligrams does 1 ml of 4°C water weigh? ____________

TEMPERATURE:

The metric measurement for temperature is the **Celsius** (formerly called the **centigrade**) scale. Like mass and volume, it is also based on water: the zero point was set at the freezing point of pure water, and the boiling point of pure water at sea level was used to set the 100° mark.

For converting between Fahrenheit and Celsius scales, use the following formulas:

°C to °F °F = (1.8 x °C) + 32° (an exact formula)
°F = 2 x °C +30 (approximate between 0–25 °C, and more easily done in your head)
°F to °C °C = (°F – 32°) / 1.8

ACTIVITY:

Your teacher has provided several items for you to guess the metric measurements of. Record your guess. Then use the appropriate measuring tool (metric ruler, graduated cylinder, or balance) to determine the actual measurement.

Item (unit)	Guess:	Actual Measurement:
1. Pour water into a mug to represent a regular-sized serving of coffee (ml)		
2. A nickel (g)		
3. A sheet of binder paper (cm^2)		
4. The height of your lab table (m)		
5. Your body temperature (°C)		
6. Room temperature (°C)		
7. A quarter (mm)		
8. Your height (m)		
9. Your weight (kg)		
10. A 1-gallon milk container (l)		
11. A paper clip (g)		
12. Distance from your home to your school (km)		Use an online mapping tool to compute exact distance.

BODY MEASUREMENTS:

It is helpful to develop a sense of the metric dimensions of objects by comparing them to known reference points on your body.

How wide (in cm) does your hand span (from the tip of the thumb to the tip of the little finger) when all your fingers are spread out completely?

When you stand, what part of your body is 1 meter above the ground? Point to where you think 1 meter is against your body. Then check your guess with a meter stick.

How much do you weigh in kilograms?

Is there any part of your hand that you could use to remind yourself of the length of 1 cm? Of 1 mm?

How wide is your armspan, in m?

You can also use your gait to help estimate distance. Walk normally on a level surface—for example, down the hallway at your school. How many steps does it take to walk 5 m?

If you record the time it takes to walk 10 m (in seconds), how long in minutes) would it take you to walk 1 km (assuming your speed did not change)?

SUMMARY TABLES OF METRIC SYSTEM UNITS

Mass and Weight:	
1 metric ton	= 1000 kg
	= 2,205 lb
1 ton	= 2000 lb
1 kilogram (1 kg)	= 2.21 lb
1 gram (1 g)	= 0.035 oz

Area:	
1 square centimeter (1 cm^2)	= 100 mm^2
	= 0.16 in^2
1 square meter (1 m^2)	= 10,000 cm^2
	= 10.76 ft^2
	= 1.19 yd^2
1 square kilometer (1 km^2)	= 1,000,000 m^2
	= 0.39 $mile^2$
	= 247.1 acres

Volume:	
1 cubic centimeter (1 cm^3)	= 1 mL
	= 0.16 in^2
1 liter (L)	= 1,000 cm^3
	= 1.06 quarts
	= 0.26 gallon
1 cubic meter (1 m^3)	= 1,000,000 cm^2
	= 1,000 liters
	= 264.2 gallons
1 cubic inch (1 in^3)	= 16.4 cubic centimeters (cm^3)
1 cubic foot (1 ft^3)	= 28.3 liters
	= 7.48 gallons

Length:	
1 centimeter (1 cm)	= 10 mm
	= 0.39 inch
1 meter (1 m)	= 100 cm
	= 39.4 inches
	= 3.28 feet
	= 1.09 yards
1 kilometer (1 km)	= 1,000 m
	= 3,280 feet
	= 0.62 miles
1 inch (1 in)	= 25.4 mm
1 foot (1 ft)	= 30.5 cm
1 yard (1 yd)	= 0.91 m
1 mile (1 mi)	= 5,280 feet
	= 1.609 kilometers
	= 0.87 nautical miles
1 nautical mile (1 nmi)	= 1.15 miles
	= 1.85 kilometers

APPENDIX B

Glossary of Marine Biology Terms

(WITH LATIN AND GREEK PREFIXES AND WORD ROOTS)

Abdomen The posterior body region of arthropods.

Abiotic (*a* = without, *bio* = life) Relating to any physical or chemical (nonliving) component in the environment, such as light, temperature, pressure, or concentration of dissolved nutrients. Opposite of **biotic**.

Aboral (*ab* = away from) Relates to the side the body of that is opposite the mouth on an animal having **radial symmetry**.

Abyssal zone The region of the open ocean between depths of 4000–6000 meters.

Accessory pigments Light-absorbing compounds in algae and other photosynthetic organisms that transfer energy to chlorophyll, assisting in the process of **photosynthesis**. **Phycobilins** and **carotenoids** are two groups.

Accretion An increase in coastal land by the gradual addition and deposition of sediments in water.

Adaptation Any physical structure or behavioral trait of an organism that helps it survive and reproduce in its environment.

Aeolid A type of sea slug that has its dorsal surface covered with fingerlike respiratory **cerata**.

Aerobic Occurring in the presence of oxygen.

Agar A polysaccharide extracted from the red alga *Chondrus crispus.* Used commercially to thicken liquids.

Agnatha (*a* = without, *gnath* = jaw) The class of jawless fish that contains lampreys and hagfish.

Alga (*pl.* **algae**) Any type of **eukaryotic**, photosynthetic **Protist**.

Algin A substance derived from the cell walls of brown algae, used commercially for thickening liquids. Also called **sodium alginate** or **alginic acid**.

Ambulacral groove A groove on the echinoderm body that houses the **tube feet**.

Amphipod (*amphi* = on both sides, *pod* = foot) A small crustacean with a laterally compressed body (its body is taller than it is wide).

Anadromous fish (*ana* = up, *dromos* = running) Marine fish that travel to freshwater habitats to spawn and reproduce; for example, salmon.

Anaerobic (*an* = without) Capable of living without oxygen.

Anoxic Completely lacking oxygen, as some mudflat sediments.

Antenna (*pl.* **antennae**) A jointed appendage (one of a pair) on the head of arthropods used for various purposes: touch (lobsters), locomotion (copepods), **suspension feeding** (mole crabs).

Anterior Toward the head region or front part of an animal. Opposite of **posterior**.

Anthozoa A class of cnidarians that includes corals and sea anemones.

Anthropogenic Related to human influence or caused by human activities.

Aperture The hole on a gastropod mollusk shell into which the animal's body can be withdrawn.

Apex The pointed tip of a spiral gastropod shell.

Aphotic zone The ocean depths where sunlight cannot penetrate.

Aquaculture The captive husbandry of economically-important marine organisms, usually for food. See **Mariculture**.

Aristotle's lantern The 5-jawed feeding apparatus of **sea urchins**.

Arrow worms Planktonic, transparent, predatory animals with grasping spines on their head region used for capturing planktonic prey.

Autotomy (*auto* = self, *otomy* = cutting) The action of voluntarily breaking off a body part or appendage. Crabs are able to autotomize a leg if one is grasped by a predator.

Autotroph (*auto* = self, *troph* = feeding) An organism that can synthesize organic material for its nutrition through **photosynthesis** or **chemosynthesis**.

Backshore The region of the beach that lies between the highest tide mark and the dune vegetation. This region is influenced by the **surf** only during severe storms.

Baleen A fibrous protein material hanging in fringes from the upper jaws of whales of the order Mysticeti. Used in feeding to strain planktonic and small nektonic organisms from seawater. Also called **whalebone**.

Basal disc The part of the sea anemone body that attaches to the **substrate**.

Batch feeding When Mysticete whales engulf large volumes of material and filter it through their **baleen** in one feeding bout. There are several methods: a large volume of water can be gulped (as **rorquals** do); benthic sediments can be strained (as grey whales do); or **neuston** can be skimmed from the sea surface (as bowhead whales do).

Bathypelagic zone (*bathy* = very deep) The region of the open ocean between depths of 1000–4000 meters.

Beach A region of **sediment** lying on the shoreline of a body of water. The beach is shaped by waves, wind, and tidal currents, which transport the sediment particles onto the **backshore**, as well as along the shore.

Beach erosion The removal of beach sand by wind, waves, tidal currents, or human activities on the shoreline.

Beach nourishment A method of temporarily counteracting beach erosion by placing huge quantities of sediments onto the beach to extend it seaward.

Beak The oldest part of a bivalve shell, synonymous with the **umbo**.

Benchmark A permanent location marker used to establish a reference elevation for a survey.

Benthic Related to the seafloor; bottom-dwelling.

Benthos A collective term for organisms that live on or in the seafloor.

Berm A feature on sandy beaches consisting of a wave-deposited sand dune with a flattened top surface that runs parallel to the shore.

Bilateral symmetry A pattern of body organization where the right and left sides of the body are externally identical to each other.

Bioaccumulation The buildup of substances within the bodies of organisms through various means of uptake, such as through food, through the skin, or across respiratory surfaces.

Bioerosion The breakdown of materials (usually calcareous matter) through the physical or chemical actions of living organisms.

Biogenic Produced by the actions or metabolic processes of living organisms.

Biological magnification A process when a persistent chemical is transferred through the bodies of animals via predation up the food chain, becoming more concentrated at higher trophic levels.

Bioluminescence The production of light by chemical means by organisms; a common feature of many types of deep-sea creatures.

Biomass The total weight of all the organisms living within a particular area.

Biotic Referring to factors originating from living organisms.

Bioturbation The movement of sediment due to burrowing and other activities of **benthic** organisms.

Bivalve Mollusks enclosed by a shell composed of 2 **valves** (i.e., mussels, clams, oysters).

Blade The portion of the seaweed **thallus** that resembles a leaf.

Bloom A sudden increase in **phytoplankton** density in response to favorable environmental conditions, such as increased nutrient levels.

Blubber An insulating layer of lipid-rich tissue beneath the epidermis of whales and other marine mammals.

BOD Biological Oxygen Demand. A measurement of water quality that estimates the amount of organic material suspended in the water.

Brachiopod A filter-feeding benthic invertebrate enclosed by two valves from which emerges a fleshy stalk. Also known as a **lamp shell**.

Brackish water Water that is intermediate in salinity between freshwater and ocean water. Brackish water ranges between 1–18 grams of salt per liter of water.

Bradycardia (*brady* = slow, *cardi* = heart) A slowing of the heart rate; occurs in marine mammals during deep diving to reduce oxygen consumption.

Broad-spectrum A substance that is effective in killing a wide range of microorganisms, both gram-positive and gram-negative bacteria. Contrast with **narrow spectrum.**

Bryozoans (phylum Ectoprocta) Tiny sessile invertebrates that reproduce clonally forming a colony; each clonal individual is called a **zooid**.

Budding A method of asexual reproduction wherein the offspring grows out from the body of the parent; occurs in lower invertebrates, such as cnidarians and sponges.

Byssus/Byssal threads Fibrous protein strands secreted by the byssal gland of certain bivalves (especially mussels), that attach the animal to a hard **substratum**.

Calcareous Containing the element calcium, or being composed of calcium carbonate.

Calcite A crystalline form of calcium carbonate found in mollusk shells.

Carapace 1) The anterior exoskeleton covering the head and thorax of some crustaceans; 2) The dorsal shell of sea turtles.

Carnassial Relating to teeth adapted for tearing and slicing flesh, such as the premolars and molars of carnivorous mammals.

Carotenoids Two types of organic pigments, the carotenes and **xanthophylls**, which range in color from reddish/orange/yellow. Functioning in photosynthetic organisms as accessory pigments and antioxidants, they are also responsible for the reddish color of tissues of certain higher consumers (such as shrimp, lobster, and salmon).

Carrageenan A substance isolated from a red alga (Irish Moss, *Chondrus crispus*) used as a thickening agent in cosmetics, foods, and industrial products.

Catadromous fish (*cata* = down, *dromos* = running) A fish that lives as an adult in freshwater, but returns to the sea to spawn, such as the American eel.

Caudal Referring to the tail or posterior end of the body.

Cellular respiration The processes by which a cell breaks down sugar or other **organic** molecules to extract energy to sustain its metabolism. Carbon dioxide is produced and released into the environment.

Cephalopods (*ceph* = head, *pod* = foot) Members of the class Cephalopoda (phylum Mollusca). Squid, octopus, *Nautilus*, cuttlefish, and argonauts.

Cephalothorax The fused head and thorax region of many types of crustaceans; most conspicuous on lobsters.

Cerata Fingerlike projections on the dorsal surface of **Aeolid** sea slugs that serve as respiratory structures for gas exchange.

Cetaceans (*ceta* = whale) Marine mammals in the order Cetacea, consisting of dolphins, porpoises, and whales.

Chaetognaths (phylum Chaetognatha *chaet* = bristle/spine, *gnath* = jaw) See **Arrow Worms**.

Chemoautotrophs Organisms such as certain bacteria and deepsea tubeworms that are capable of **chemosynthesis**.

Chemoreception The ability to sense chemicals. Marine animals have a variety of structures to detect chemicals dissolved in seawater.

Chemosynthesis A method of manufacturing high-energy compounds by oxidizing inorganic materials in the absence of sunlight.

Chitin A durable polysaccharide that makes up the arthropod **exoskeleton**.

Chiton An intertidal, **grazing** mollusk with 8 hinged shell plates that provide protection, but allow flexibility.

Chlorophyll (*chlor* = green, *phyl* = leaf) A green pigment that absorbs sunlight, providing energy needed in the process of **photosynthesis** to manufacture carbohydrates from carbon dioxide and water.

Choanocyte (*choan* = funnel, *cyt* = cell) A special flagellated cell in sponges that creates a water current for filter feeding. Also called a **collar cell**.

Chondrichthyes (*chond* = cartilage, *ichth* = fish) A class of vertebrates comprising fish with skeletons made of cartilage: sharks, rays and skates, and chimeras.

Chromatophore (*chrom* = color, *phor* = carry) A pigment-containing cell on the outer surface of the bodies of fish and cephalopods. The pigment is condensed or expanded within the cytoplasm to lighten or darken the coloration and to change the color pattern of the animal.

Ciguatera A human disease caused by eating a tropical fish having **dinoflagellate** toxins concentrated within its muscle tissue. Symptoms include nausea, diarrhea, numbness, and hallucinations.

Cilia (*cilium* = eyelash) Tiny hairlike structures that beat from side to side on the cell surface of multicellular and unicellular organisms that serve to produce a current or to locomote the organism.

Cirri (*sing.* **cirrus**) The feathery appendages of barnacles that capture plankton from the water column.

Claspers A pair of external structures located along the pelvic fins of male cartilaginous fish, used for internal fertilization of the female.

Clone An organism that has the same genetic makeup (i.e., has identical DNA) as its parent.

Cnidaria (*cnid* = nettle) A phylum comprising 3 classes of sessile or pelagic animals that have stinging tentacles.

Cnidoblast (*cnid* = nettle, *blast* = a growing cell) A specialized cell (unique to members of the phylum Cnidaria) that creates a stinging microstructure called a **nematocyst**.

Coccolithophore (*cocco* = berry, *lith* = stone, *phor* = to carry) A unicellular, **flagellated** alga with a cell surface covered with numerous tiny **calcareous** plates.

Collar cell See **Choanocyte**.

Com- A prefix meaning *together*.

Comb jellies See **Ctenophores**.

Commensal species An animal that lives on, in, or with the **host species**, neither helping nor harming the host.

Commensalism A type of **symbiosis** wherein two animal species (the **commensal** and the **host**) live and interact in close proximity together without apparent competition.

Community A group of populations of different species living within a particular habitat.

Conceptacle A cavity on the **thallus** of a seaweed housing the reproductive tissues.

Contractile vacuole An **organelle** in ciliates used for maintaining water balance; it periodically expels water diffusing into the cell, preventing the cell from bursting.

Copepods (*cope* = oar, *pod* = foot) Small herbivorous, planktonic crustaceans found throughout the world ocean. The most abundant type of animal on our planet.

Coralline algae Certain types of red and green algae that produce calcium carbonate within their cells.

Cordgrass See ***Spartina***.

Cosmopolitan species A species that is found throughout the world wherever the habitat is suitable. Marine examples: Many types of cetaceans, the Great white shark, the moon jelly *Aurelia aurita*.

Countershading A coloration pattern consisting of a dark upper body surface and a light ventral surface, common to most **nektonic** organisms in the **epipelagic zone**. This pattern helps the organism blend in with the background when viewed either from above or from below in the water column, making the organism less visible to predators.

Crinoids Suspension-feeding echinoderms with numerous feathery arms. There are free-living forms called feather stars, and stalked forms, called sea lillies.

Crustaceans (*crusta* = shell) A large category of arthropods that includes the barnacles, shrimps, lobsters, crabs, **copepods**, **amphipods**, and **isopods**.

Ctenophores (*ctene* = comb, *phor* = to carry) Members of the phylum Ctenophora. Gelatinous planktonic predators that swim with rows of tiny beating combs. Also called **comb jellies**.

Cypris The final stage of barnacle larval development following the planktonic **nauplius** stage. The cypris settles on a hard **substrate** and undergoes **metamorphosis** into the benthic adult form.

Decapods (*deca* = ten, *pod* = foot) Ten-legged crustaceans in the order Decapoda. Includes shrimp, lobsters, and crabs.

Decomposer An organism that breaks dead organic matter into inorganic materials, releasing nutrients into the ecosystem. Examples: bacteria and fungi.

Demersal Related to the seafloor; living near the sea bottom.

Density 1) The mass of a substance per unit volume; 2) The number of organisms per unit area.

Deposit feeder An animal that obtains its food by ingesting benthic sediments and digesting the organic matter contained within as worms, heart urchins, and sea cucumbers do.

Desiccation The process that occurs when the tissues of organisms dry out through exposure.

Detritus Decomposed organic matter, and food for numerous types of **deposit feeders**.

Diatoms Photosynthetic unicellular organisms encased by a glass skeleton. There are planktonic, as well as benthic, species; both are important **primary producers** in the marine ecosystem.

Dichotomous Divided into two parts or branches, as the **stipes** of various seaweeds.

Diel vertical migration (DVM) A phenomenon occurring throughout the world ocean when **zooplankton** from lower depths swim up into the **epipelagic zone** at twilight to feed on **phytoplankton** and other zooplankton, returning to the depths before sunrise.

Diffusion A passive, physical process that occurs when substances move from a region of high concentration to a region of low concentration. See **Osmosis**.

Dinoflagellates Unicellular planktonic organisms (usually photosynthetic) that use two flagella for locomotion.

Diploid Refers to a cell that contains two copies of each chromosome normally found in the species.

Disruptive coloration A color pattern that breaks up the outline of an animal's body, making it less noticeable to predators.

Diurnal tide A tide cycle consisting of one high and one low tide per day.

Diversity The total number of species that inhabit a designated area.

DOM Stands for Dissolved Organic Matter. Refers to measurements of water quality.

Dorid A type of sea slug with respiratory structures clustered together on its dorsal surface. See **Aeolid**.

Dorsal Related to the upper body surface of an animal; the opposite of **ventral**.

Downwelling The vertical movement of surface water downward into deeper water.

Dulse A type of edible red algae (*Palmaria palmate*) found on both sides of the North Atlantic.

Dune A pile of windblown sand on the **backshore**, shaped like a small hill.

Dune crest The highest point of a sand dune.

Dune slope A side of a sand dune that descends from the dune crest down toward the water.

Echinoids Animals in the phylum Echinodermata that have a rounded or flattened calcareous **test** covered with spines, and a 5-toothed feeding structure. Sea urchins, heart urchins, sand dollars, and sea biscuits.

Echolocation The process of determining the position and identity of underwater objects by emitting sounds, then analyzing the echoes that are reflected by the targeted objects. Marine mammals, bats, and some birds use this ability to hunt and navigate in low-vision environments.

Ecology (*eco* = home, *logy* = study) The study of the abundance and distribution of organisms, and their interrelationships with their environment.

Ecosystem The physical environment of a particular area (including all the **abiotic** factors) together with all the populations of different species living within the area.

Ecto- A prefix meaning *external, outside, out,* or *on the surface* (fleas are ectoparasites on dogs).

Ectotherm An organism whose internal body temperature is dependent on the external ambient temperature.

Ectothermic Unable to regulate the internal body temperature through physiological means. Marine invertebrates and the majority of fish species are **ectotherms**.

Elasmobranchii A subclass of **Chondrichthyes** that contains the sharks, rays, and skates.

Encrusting organism One that grows over a hard surface, covering it with a sheet or crust of living tissue.

Endemic Native to a specific location, and not found anywhere else.

Endo- A prefix meaning *internal, inside, in* or *within the surface* (tapeworms are endoparasites).

Endocrine disruptor A substance that alters the normal functioning of the glands in the endocrine system, causing adverse effects to the organism, its offspring, or its population.

Endothermic Able to regulate the internal body temperature and maintain it at a constant level through physiological means; **endothermy** is a characteristic of mammals and birds.

Epi- A prefix meaning *on* or *upon a surface.*

Epifauna Animals that live on a firm substrate or on the surface of other animals.

Epipelagic zone The region of the open ocean from the surface extending to a depth of 200 meters; it includes the **photic zone**.

Epiphytes (*phyt* = plant) Algae that grow on the surface of other algae.

Epitheca (*theca* = case) One of the two valves that form a diatom **frustule**. The epitheca is larger than the lower valve (the **hypotheca**).

Estuary A wetland area where fresh- and saltwater meet, such as at the mouth of a river (*Adj.* **estuarine**).

Eukaryotic Referring to cells that have organelles, a nucleus, and linear DNA.

Euryhaline (*eury* = wide, *hali* = salt) Capable of surviving under a wide variety of salinity conditions: in either salt-, fresh-, or brackish water. Many **estuarine** species are euryhaline.

Evolution A change in the frequency of inherited traits between successive generations of a population.

Exoskeleton A protective material made of **chitin** that supports the bodies of arthropods.

Feather stars See **Crinoids**.

Filter feeder A sessile animal that obtains food by creating a water current through its body and removing any suspended organic particles. Examples are clams, **sea squirts**, and sponges.

Fission A method of asexual reproduction where one organism splits into two genetically identical individuals, or **clones**.

Flagellum (*pl.* **flagella)** A long whip-like strand on the surface of certain cells, used for cell locomotion (e.g., sperm cells).

Flipper The anterior appendage of a marine mammal or sea turtle.

Fluke The horizontally flattened, rear appendage of **Cetaceans,** used for forward propulsion. Flukes move in an up-and-down direction, as opposed to the side-to-side movement of the fish **caudal** fin.

Food chain A representation of a linear transfer of energy from primary producers to higher consumers, showing what each species eats.

Food web A representation of the combined feeding relationships existing between multiple species within a particular community.

Foot A large muscle in mollusks that has been modified in several ways: for burrowing in **bivalves**, for crawling in **chitons** and **gastropods**, and for grasping prey in **cephalopods**.

Foraminiferans (*foram* = a small opening, *fer* = to bear or carry) Unicellular **protists** covered with a perforated **calcareous** covering. There are planktonic and benthic forms.

Foreshore The intertidal region that lies between the average low tide mark and the average high tide mark. See **Backshore**.

Fouling community A group of **epifaunal** organisms typically found on the surface of continuously submerged structures, such as ship hulls and pier pilings.

Frustule The siliceous skeleton of a diatom; it consists of an **epitheca** and a **hypotheca**.

Fucoxanthin An accessory pigment found in brown algae; it functions to assist in the process of **photosynthesis** by absorbing light rays in the blue-green and yellow-green part of the spectrum.

Gametes **Haploid** reproductive cells that unite to form a **zygote** (i.e., egg and sperm cells).

Gastropods (*gaster* = stomach, *pod* = foot) Mollusks that are members of the class Gastropoda. Sea slugs and marine snails.

Genus (*pl.* **genera**) A taxonomic group above the rank of species. The genus is the first part of an organism's scientific name, as the genus *Homo* in *Homo sapiens*.

Gill arch A bony or cartilaginous curved structure that supports the gills of fishes.

Gill filaments Thin, feathery projections where gas exchange occurs in fish; they are connected to the **gill arch**.

Gill rakers Forward-pointing projections on the **gill arches** of fish that prevent prey from entering the gill chamber and protect the gills from damage when prey is swallowed. **Planktivorous** fishes such as anchovies have long, narrow gill rakers to help strain food particles from the water.

Girdle The flexible tissue surrounding the shell plates of a **chiton**.

Glassy layer Part of the sandy beach **foreshore** that remains constantly saturated with seawater.

Gonad A reproductive organ that creates the **gametes** (can refer to either a male or female organ).

Gorgonian A member of the phylum Cnidaria, class Anthozoa; a soft coral with a flexible, internal skeleton. Sea fans and sea whips.

Grazer An animal that obtains food by scraping attached organisms (usually algae) off hard **substrates**.

Habitat The environment where an organism lives.

Hadal zone The deepest areas of the ocean, below 6,000 m; characteristic of ocean trenches.

Halophyte (*hali* = salt, *phyt* = plant) An intertidal or **estuarine** plant that is tolerant of saline water and soils with high salt concentrations.

Haploid Refers to a cell that contains only half the amount of genetic material that is usually found in cells of that species (i.e., contains only one copy of each chromosome). See **Diploid**.

Harem A group of female marine mammals guarded by a dominant male to maintain his exclusive reproductive access. Harems are common in **pinnipeds**.

Harmful algal bloom (HAB) A **phytoplankton bloom** that produces toxins or causes some negative effect on the coastal community of marine organisms. Formerly called a **red tide**, but many kinds of harmful blooms do not color the water red, so HAB is the preferred term.

Hemocyanin A copper-containing pigment that carries oxygen through the bloodstream of arthropods and mollusks.

Herbivore An organism that obtains its nutrition by consuming **autotrophic** organisms (*Adj.* **herbivorous**).

Hermaphrodite An organism that possesses both male and female reproductive organs. Nudibranchs, flatworms, and barnacles are examples of **hermaphroditic** species.

Heterocercal (*hetero* = different, *cercal* = tail) Refers to the asymmetric caudal fin of sturgeons and many types of sharks. The upper and lower lobes of the fin are unequal in size. See **Homocercal**.

Heterodont Referring to unequally-shaped teeth (a feature of most mammals). Contrast with **homodont.**

Heterotroph (*hetero* = different, *troph* = feeding) An organism not capable of conducting **photosynthesis** or **chemosynthesis** and so must obtain its energy by consuming organic materials produced by other organisms.

Holdfast A portion of the algal **thallus** that anchors the alga to the substrate.

Holoplankton (*holo* = complete) Organisms that spend their entire life cycle as members of the planktonic community.

Holothuroids Members of the phylum Echinodermata with elongated bodies and tentacles; namely, the sea cucumbers.

Homeostasis (*homeo* = same) The self-maintenance of constant internal conditions within an organism's body by monitoring and adjusting physiological processes.

Homeotherm An organism that regulates its body temperature; body temperature is maintained at a constant level independent of the temperature of the environment. See **Poikilotherm**.

Homocercal (*homo* = same, *cercal* = tail) Refers to the symmetric caudal fin of most bony fish and fast-swimming pelagic sharks. The upper and lower lobes of the fin are equal in size. See **Heterocercal**.

Homodont Referring to teeth of similar size and shape (a feature of reptiles, fish, and odontocetes). Contrast with **heterodont.**

Host species A species that provides a home for another species.

Hurricane A large-scale tropical storm with winds blowing above 74 miles per hour. Storm winds blow in a counterclockwise direction in the northern hemisphere and clockwise in the southern hemisphere. See **Typhoon**.

Hydroid A type of colonial Cnidarian (class Hydrozoa) that has the **polyp** form as the dominant stage of the life cycle.

Hypersaline (*hyper* = greater) Refers to water with salinity greater than 40 ppt (> 40 grams of dissolved salt per liter).

Hypotheca (*hypo* = below) One of the **valves** of a diatom **frustule**; it is smaller in diameter and fits within the larger **epitheca**.

Hypoxia A condition of very low oxygen levels.

Ichthyology (*ichth* = fish, *logy* = study) The study of bony and cartilaginous fishes.

Infauna Marine animals that live in or within sediments or hard substrata.

Interdigitate To interlock together like fingers do when both hands are put together. The term describes the upper and lower teeth of cetaceans, when the mouth closes.

Interstices Spaces between sediment particles; these spaces may contain air or water (*Adj.* **interstitial**).

Intertidal zone The area of shoreline bounded by the high and low water marks.

Invertebrate Any type of animal that lacks a backbone.

Iridophore (*irid* = rainbow, *phor* = to carry) A specialized skin cell that contains reflective crystals of the amino acid guanine, producing an iridescent sheen. Common in the epidermal cells of silvery fish. Also called **iridocyte**.

Isopod (*iso* = equal, *pod* = foot) benthic crustacean with seven pairs of legs and a body flattened from top to bottom.

Iteroparous Reproducing more than once during the life cycle.

Kelp Various species of large brown algae that belong to the order Laminariales.

Krill Small pelagic shrimp-like crustaceans (order Euphausiacea) found in Arctic and Antarctic waters; these primary consumers are important food sources for **Mysticete** whales.

Lamp shells See **Brachiopods**.

Lancelets Bilaterally symmetric, cigar-shaped, filter-feeding invertebrates in the class Cephalochordata.

Larva An early life stage of a marine animal that undergoes **metamorphosis** into the adult form.

Larvaceans **Holoplanktonic** animals in the phylum Chordata that construct houses of mucus and filter feed on **nanoplankton**.

Lateral line A system of pores running along the sides of fishes, used for detecting water currents, pressure, and vibrations.

Lecithotrophic larva (*lecitho* = yolk, *troph* = feeding) A planktonic larva that does not feed on other plankton, but is nourished by the yolk material from the egg from which it hatched.

Leeward The direction *toward* which the prevailing wind is blowing. On a beach, windblown sand is deposited on the leeward side of shells, seaweed piles, or other objects.

Leptocephalus larva (*lepto* = slender, *ceph* = head) The elongated, flattened transparent larva of eels.

Light organ See **Photophore**.

Limpet A type of intertidal gastropod with a conical shell.

Littoral zone Synonym for **intertidal zone**.

Lorica A protective casing that encloses certain kinds of unicellular **protists**.

Macrofauna Marine animals larger than 1.0 mm in size.

Macroplankton Planktonic organisms between 2–20 cm in size.

Madreporite A **calcareous** plate on the **aboral** surface of sea stars and sea urchins that takes in water for the **water vascular system**. See **Sieve Plate**.

Mammology The scientific study of mammals.

Mandible The bone of the lower jaw, in chordates. Also a mouthpart for seizing and biting food in marine arthropods and terrestrial insects.

Mantle In mollusks, a tissue that secretes the shell or forms a cavity that encloses the gills.

Mantle cavity A chamber housing the mollusk gills, it is formed by mantle tissue.

Mariculture The process of culturing marine organisms in their natural habitats. See **Aquaculture**.

Marine snow A fluffy aggregated substance composed of dead organic material, bacteria, fecal particles, and inorganic debris. It forms in the upper **epipelagic zone** and continually drifts downward toward the sea floor, where it is consumed by zooplankton, fish and benthic invertebrates.

Maxilliped A modified appendage on the thorax of marine arthropods, used for creating a water current for filter feeding and filtering water as it passes through to the gills.

Medusa A planktonic body form of Cnidarians, consisting of a downward-directed mouth and dangling tentacles, as in jellyfish.

Megalops The last planktonic larval stage of crabs before **metamorphosis** transforms the larva into the benthic adult form.

Megaplankton Plankton organisms between 20–200 cm in size, such as **Salps** and **Siphonophores**.

Meiofauna Animals between 0.1–1.0 mm in size; usually found living between sand grains. Also called **interstitial** fauna (see **Interstices**).

Melon An enlarged fatty organ on the forehead of most toothed whales (order **Odontoceti**). It is used to focus sound waves produced by the animal in the process of **echolocation.** The sperm whale has the largest melon of all whales.

Meroplankton (*mero* = part, partial) Organisms that spend the early part of their life cycle passively drifting as members of the **plankton** community, then undergo metamorphosis into a **benthic** (bottom-dwelling) or **nektonic** (actively swimming) adult form.

Meso- Prefix meaning *middle.*

Mesoglea The jellylike material found between the ectodermis and endodermis of Cnidarians.

Mesopelagic zone The region of the open ocean between depths of 200–1000 meters; it includes the **oxygen minimum zone.**

Mesoplankton Planktonic organisms between 0.2–20 mm in size, such as copepods and most invertebrate larvae.

Metamorphosis The process in which the body of a larva is reorganized and transformed into a different shape.

Microfauna Animals smaller than 0.1 mm.

Microplankton Planktonic organisms between 20–200 μm in size. Examples include **Diatoms**, **dinoflagellates**, and **tintinnids**.

Mixed semidiurnal tide When there are 2 high and 2 low tides each day; both high tides and both low tides are different in height. See **Semidiurnal tide.**

Molt An arthropod exoskeleton that has been cast off during the **molting** process.

Molting The process of shedding an external covering (such as feathers, fur, or an **exoskeleton**) and replacing it with new growth.

Morphology The form and structure of a living organism or the shape of any of its parts; also, the scientific study of the form and structure of organisms.

Mutualism A type of **symbiosis** wherein two organism live in close proximity and confer benefits to each other from their respective activities. See **parasitism, commensalism.**

Myoglobin An iron-containing molecule that stores oxygen in muscle tissue.

Mysticetes (*myst* = moustache) Baleen whales (order Cetacea, family Mysticeti).

Nacre The shiny mother-of-pearl material that lines the inside of mollusk shells, especially noticeable in abalone.

Nanoplankton Planktonic organisms that are between 2–20 micrometers in size; e.g., **coccolithophores.**

Narrow-spectrum A substance that is effective in killing only a limited variety of microorganisms. Contrast with **broad spectrum.**

Natural selection The process by which favorable inherited traits become more common in successive generations of a population, while inherited traits that are not helpful to the survival and reproduction of the species become less common.

Nauplius A common type of coastal **zooplankton.** It is the first larval stage of barnacles and certain microcrustaceans.

Neap tide A tide of decreased range occurring at the first or third quarter of the lunar month, when the moon is 1/4 or 3/4 full.

Nekton Animals that have the ability to swim throughout the oceans, independent of the direction of waves and currents.

Nematocyst (*nema* = thread, *cyst* = capsule) A specialized stinging apparatus housed within the **cnidoblasts** of Cnidarians, used for prey capture.

Neritic zone The area of coastal ocean that overlies the continental shelf.

Net plankton The planktonic organisms that can be collected using a standard plankton net (i.e., a net with a mesh size greater than 0.2 mm).

Neurotoxin (*neur* = nerve) A poison that directly affects the nervous system.

Neuston Planktonic organisms that live at the air/water interface of the ocean surface.

Niche The role of an organism in its ecosystem, including its trophic relationships with other species and how it uses the physical environment.

Nitrate (NO_3^{-1}) An important nutrient necessary for the growth of **phytoplankton.**

Nori The Japanese term for species of the red alga genus *Porphya* that are used to wrap rice and seafood in the preparation of sushi.

Notochord A dorsal rodlike support tissue present in animals of the phylum Chordata.

Nudibranch (*nudi* = naked, *branch* = gill) Marine gastropod mollusks that lack a shell; also called sea slugs.

Nutrient A type of inorganic compound needed by primary producers for nutrition and growth. **Nitrate** and **Phosphate** are two important kinds.

Odontoceti (*odont* = teeth, *ceta* = whale) The family of toothed whales; dolphins and porpoises within the Order **Cetacea.**

Oligo- Prefix meaning few or several. See **oligochaete, oligotrophic.**

Operculum 1) A bony plate protecting the gills of fish; 2) A calcareous or protein structure that covers the shell **aperture** of gastropod mollusks, protecting the body after it is withdrawn into the shell.

Ophiuroids (*ophi* = snake) Members of the class Ophiuroidea, phylum Echinodermata. Brittle stars.

Organelle Within a cell, a membrane-bound compartment that performs a specialized function and encloses a chemical environment different from the cytoplasm.

Organic Refers to any substance that contains the element carbon.

Organism Any living entity.

Ornithology (*ornith* = bird, *logy* = study) The scientific study of birds.

Osculum (*pl.* **oscula)** (*os* = mouth) An opening on the body of sponges where excurrent water is expelled.

Osmo A prefix referring to *osmosis/water balance.*

Osmoconformer An organism whose internal salt concentration fluctuates with the changing salinity of the surrounding water.

Osmoregulation The process of actively regulating the concentration of dissolved substances within body tissues to keep the levels constant over time; that is, to maintain **homeostasis.**

Osmoregulator An organism that can maintain a cellular concentration of dissolved substances either higher or lower than the salinity of the surrounding water.

Osmosis The passive movement of water across a cell membrane into an area of higher salinity or higher solute concentration. Osmosis is a special case of **diffusion**.

Osteichthyes (*osteo* = bone, *ichth* = fish) The class of fish in phylum Chordata whose members have skeletons made of bone.

Ostia (*sing.* **ostium)** Numerous tiny holes on the body of a sponge where water is taken in. Also, perforations in the heart of crabs.

Otoliths (*oto* = ear, *lith* = stone) Calcareous structures in the inner ear of fish, used for balance and hearing.

Ovipary (*ovo* = egg) Reproduction by producing eggs that are externally fertilized and develop outside the mother's body (*Adj.* **oviparous**).

Ovovivipary (*ovo* = egg, *viv* = life) Reproduction by producing eggs that are fertilized internally and retained in the mother's body until time of hatching. Developing embryos within the eggs do not receive nourishment from the mother (*Adj.* **ovoviviparous**).

Oxygen minimum zone The region between 200–1000 meters where the concentration of dissolved oxygen is lowest. Here **photosynthesis** cannot occur to replenish oxygen consumed by animal respiration and decomposition.

Pan- Prefix meaning all: Reef-building corals have a pan-tropical distribution pattern; while copepods have a pan-oceanic distribution.

Paralytic shellfish poisoning One type of shellfish poisoning due to the effects of **saxitoxin**, a potent neurotoxin produced by certain **dinoflagellates** and concentrated within the tissues of filter-feeding **bivalves**. When a human eats a clam, mussel, or other bivalve harboring the toxin, nausea, diarrhea, loss of muscle coordination, and death may occur.

Parapodium (*para* = beside, *pod* = foot) A fleshy projection on both sides of each body segment of **polychaete** worms; often bearing bristles. Plural: **parapodia.**

Parasitism A type of **symbiosis** when one animal (the parasite) lives on or in the body of another and derives nourishment from the **host species.**

Pectoral fin A fin found just behind the gill opening on both sides of the fish body.

Pedicellariae A tiny pincher-like structure on the surface of sea stars and sea urchins, used for cleaning and defense.

Pelagic Relating to the open ocean (the region of ocean beyond the continental shelf).

Pelvic fin A fin found on the ventral surface of the fish body, toward the head.

Pen A clear, thin, quill-like structure that provides some rigidity to the body of a squid. Also called the gladius.

Pentamerous Made up of 5 parts or divisions. A characteristic of echinoderm anatomy.

Periostracum (*peri* = around, *ostrac* = shell) The outer layer of marine mollusk shells; it is usually a protein-based material covering the exterior surface of the shell.

Pharmacology The study of the sources, actions, and properties of drugs, and the body's reaction to drugs.

Phosphate (PO_4^{-3}) An important **nutrient** used by **phytoplankton** to synthesize DNA and other key biological molecules.

Photic zone The depth of ocean where enough light can penetrate to allow **photosynthesis** to occur. Usually the photic zone extends to 200 m.

Photophore (*photo* = light, *phor* = carry) A complex type of organ that produces **bioluminescence**. Also called **light organ**.

Phycobilins Light-capturing molecules found in certain types of algae that transfer energy to **chlorophyll** and in so doing, increase the efficiency of **photosynthesis**.

Phycocolloids Substances extracted from the cell walls of certain red and brown algae used for thickening, emulsifying, and gelling liquids. **Agar** and **algin** are commercially important phycocolloids.

Phycology (*phyco* = seaweed, *logy* = study) The scientific study of algae and seaweeds.

Phylogeny The evolutionary history and development of a particular group of organisms or **taxon**.

Phylum (*pl.* phyla) A taxonomic rank that is below the level of Kingdom and above the level of Class in biological classification. Plural: phyla.

Phytoplankton (*phyt* = plant) **Autotrophic** members of the plankton community.

Picoplankton Plankton that range between 0.2–2.0 μm in size, such as bacteria.

Pinnipeds (*pinna* = feather, *ped* = foot) Aquatic mammals in the suborder Pinnipedia. True seals, fur seals, sea lions, and the walrus.

Planktivorous Refers to organisms that consume plankton in their diet.

Plankton Passively drifting organisms that are transported by waves, tides, and ocean currents.

Planktotrophic larva (*troph* = feeding) A planktonic larva developing from a tiny egg that does not provide enough food energy to fuel the larva's development into a juvenile form; thus, the larva must eat other plankton to continue its growth and development. See **lecithotrophic larva**.

Planula (*pl.* planulae) The ciliated planktonic larva of **Cnidarians**.

Pluteus A larval form of sea urchins and brittle stars.

Pneumatocyst (*pneumo* = air, *cyst* = capsule) A gas-filled structure in brown algae that provides positive buoyancy to the **thallus**, orienting it toward the surface and sunlight.

Pneumatophore A gas-filled bladder used for floatation by the Portuguese man-o-war and other colonial hydrozoans. Also a type of plant root specialized for obtaining oxygen in the anoxic conditions of an estuary.

Poikilotherm (*poikilo* = variable) An organism that is unable to regulate its body temperature through physiological means; instead, its body temperature reflects the temperature of the environment. See **Homeotherm**.

Polychaetes (*poly* = many, *chaete* = bristle/spine) Marine worms in the class Polychaeta (phylum Annelida). There are swimming, crawling, and **sessile** species.

Polyp A body form characteristic of Cnidarians consisting of a **sessile** column with a central mouth ringed with tentacles that extend upward into the water column.

Porifera (*por* = hole, *fer* = to carry) The phylum of the simplest, most primitive multicellular animals: the sponges.

Portuguese man-o-war A planktonic colonial hydrozoan that floats on the surface of warm oceans, preying on fish that become entangled in its long, stinging tentacles.

Posterior Relating to a position near the rear or toward the tail end of an animal. See **Anterior**.

Preening Grooming activities of birds, such as fluffing and tidying feathers, etc.

Primary consumer An organism that consumes a **primary producer**.

Primary producer An autotrophic organism that produces energy-rich organic materials through the process of **photosynthesis** or **chemosynthesis**.

Primary production The production of high-energy organic compounds from carbon dioxide and water through **photosynthesis** or **chemosynthesis**.

Productivity The rate at which photosynthetic products (carbohydrates and other organic compounds) are formed from inorganic carbon sources (i.e., from CO_2).

Profile A plotted line that illustrates the changes in elevation of a landform.

Prokaryote An organism with cells that lack **organelles** and a nucleus, such as bacteria and blue-green algae.

Protists **Eukaryotic** organisms that are not classified as animals, plants, fungi, or bacteria. There are unicellular and multicellular species, including **diatoms**, **tintinnids**, **protozoans**, and seaweeds.

Protozoans (*proto* = first, *zoa* = animal) Unicellular aquatic organisms that feed on **detritus**, bacteria, algae, and other protozoans, though some kinds can also can also be **autotrophic** and conduct **photosynthesis**.

Pteropods (*ptero* = wing, *pod* = foot) **Holoplanktonic** mollusks that swim by flapping two sheets of tissue, giving them the common name of sea butterflies.

Quartz One of the most common minerals found in beach sand. It is formed from igneous processes.

Radial symmetry A type of body organization where all the body parts are arranged symmetrically around a central point, like spokes on the hub of a wheel.

Radiolarians Unicellular planktonic protozoans with a glassy, spiny skeleton.

Radula A toothed, rasping feeding structure found only in the phylum Mollusca. Some mollusks use their radula for **grazing** algae; predatory gastropods have radulas modified for harpooning prey and injecting poison. The radula of cephalopods is used to drill holes through the hard body covering of their prey.

Raphe A groove that runs lengthwise down the middle of the valves of pennate diatoms.

Ray A cartilaginous fish with a flattened body and enlarged **pectoral fins** that undulate for propulsion. Most kinds are **benthic** in lifestyle, but the manta ray is **nektonic**.

Red tide A phenomenon when certain marine microorganisms **bloom** (suddenly become extremely abundant in the water column), and the dense concentration of cells colors the seawater a reddish or brown shade. Toxins released by some types of red tide organisms can kill fish and accumulate within the tissues of filter-feeding bivalves. Upon their consumption by humans, death from **Paralytic Shellfish Poisoning** may occur. See **Harmful Algal Bloom**.

Rhinophores (*rhin* = nose, *phor* = carry) Dual projections on the head region of sea slugs, used for **chemoreception**.

Ribbon worms Unsegmented worms in the phylum Nemertea. There are **benthic** (bottom-dwelling) and **pelagic** (open-ocean) forms.

Rockweeds Bushy intertidal brown algae found on temperate rocky shorelines.

Roe The unspawned egg masses of fish and invertebrates.

Rorquals The group of **Mysticete** whales with long grooves running below the mouth that allow the animal to take in immense quantities of water when **batch feeding**. Examples are the humpback whale, blue whale, and minke whale.

Salinity The amount of inorganic minerals dissolved in water, commonly expressed as total grams of mineral salts per kilogram of water.

Salps Planktonic solitary or colonial filter-feeding chordates (class Thaliacea), with gelatinous bodies.

Salt gland A specialized gland that concentrates and excretes salt from the bodies of marine birds and reptiles.

Salt marsh A temperate zone, intertidal region of coastline characterized by **brackish** water, muddy and silty sediment deposits, and a vegetation community based on **halophytes**. See ***Spartina***.

Sand Sediments ranging in size from 0.06 mm to 2.0 mm; can consist of particles eroded from the land and transported to the coastline by rivers, glaciers, or wind; or particles can consist of the remains of calcareous organisms (i.e., seashells, coral skeletons).

Saxitoxin A **neurotoxin** produced by certain types of **dinoflagellates**.

Scavenger An animal that feeds on dead organisms.

Schooling A common behavior in fish where individuals swim closely together as a group, maintaining a high degree of cohesion and coordination.

Scud A type of amphipod crustacean found in beach wrack and tidepools.

Sea squirts Sessile solitary or colonial filter-feeding chordates (Class Ascidiacea); also known as **tunicates** and **ascidians**.

Sediment Any type of particulate material transported and deposited by water. Sediments such as sand, silt, and clay, cover the ocean floor and provide a **habitat** for **infaunal** animals.

Semelparous Reproducing once during the life cycle.

Semidiurnal tide Two high tides and two low tides occurring in each daily tidal cycle. Both the high tides are approximately the same height, as are the two low tides. See **Mixed semidiurnal tide**.

Sessile Attached to the **substrate**; nonmoving, immobile.

Setae 1) Bristles on the annelid body, attached to **parapodia**; 2) Bristles on the **exoskeleton** of marine arthropods.

Sieve plate See **Madreporite**.

Silica Silicon dioxide (SiO_2); a major component of the skeletons of **diatoms**, **radiolarians**, and some sponges.

Siliceous Containing the compound silica.

Siphon An extension of the **mantle** tissue in mollusks, used to direct expelled water for jet propulsion (in squid), or for water intake (in bivalves and gastropods).

Siphonal notch An extension of the gastropod **aperture** that surrounds the tubular **siphon**.

Siphonophores Large pelagic, planktonic Cnidarians that live as clonal colonies composed of numerous specialized polyps. See **Portuguese Man-o-war**.

Sirenia An order within the Class Mammalia. A group of four species of herbivorous marine mammals found in tropical regions in North America, South America, West Africa, and the Coral Sea. Manatees and dugong.

Spartina A genus containing 14 species of salt-tolerant coastal grasses that grow in marshes and sand dunes in temperate zones. Also called **cordgrass**.

Spawning When **gametes** are released into the water column by organisms of the same species. The gametes unite in the process of external fertilization. Most fish and invertebrates **spawn** at a certain time of year.

Species A **taxonomic category** in which all the members are capable of interbreeding and producing fertile offspring.

Specific gravity The ratio of the density of a given volume of a substance to the density of an equal volume of water. Gold has a specific gravity of 19.3; i.e., a given volume of gold weighs 19.3 times more than the same volume of water.

Spermatophore A mass of material that contains sperm cells and is transferred to the female by the male during mating. Male crustaceans and cephalopods produce these.

Spicule A calcareous or siliceous supporting element, usually microscopic, in the sponge body.

Spiracle In benthic sharks and rays, a muscular opening found **posterior** to the eye, and connected to the gill chamber. Water is pumped in by opening and closing the spiracle, and flows through the gills.

Spiral valve A coiled structure within the intestines of sharks that increases the amount of surface area available for digestion and nutrient absorption.

Spongin A protein that forms a supportive material throughout the body of certain types of sponges.

Spring tide A tide that occurs at or near the time of new or full moon; a spring tide provides the highest high tides and lowest low tides of the month.

Stations Locations where data is collected along a **transect**.

Stenohaline (*sten* = narrow, *hali* = salt) Referring to organisms that cannot survive under widely fluctuating salinity conditions. These organisms have a narrow range of tolerance for salinity.
Stipe The part of the seaweed **thallus** that corresponds to a plant stalk or stem.
Strand line A line of debris cast shore onto the beach during a previous high tide. It can contain seaweed, trash, driftwood, and other floating materials. Also called the **wrack line**.
Structural color An iridescent color produced by the scattering of certain wavelengths of light off a material, in contrast to a color produced by a pigment. The luminous shine of peacock feathers and mother-of-pearl is due to light-scattering microstructures.
Sublittoral (*sub* = under, beneath) The nearshore region below the lowest part of the intertidal zone that is never exposed by a low tide. Also called the **subtidal zone**.
Substrate Any solid material that an organism can live in or on, for example: sand, mud, rock, or the body surface of another animal.
Supralittoral (*super* = above) The region of shoreline that is above the reach of the highest tide, but receives moisture from salt spray. Also called the **splash zone**.
Surf The waves that break onto a shore.
Suspension feeder An animal that captures drifting food particles from the water column. Hard corals, sea fans, and feather duster worms are examples.
Swash lines Markings on the **foreshore** of a sandy beach made by the surf, showing where the waves have reached.
Swim bladder A gas-filled organ within the body cavity of a bony fish, used to regulate buoyancy. Also called **gas bladder**.
Swimmerettes Appendages on the **abdomen** of **decapod** crustaceans, used for swimming and holding eggs.
Symbiosis (*sym* = together, *bio* = life) A phenomenon wherein two species associate and interact in close proximity with one another.
Taxon (*pl.* **taxa)** Any type of taxonomic category or taxonomic unit, such as Kingdom, Phylum, Class, Order, Family, Genus, or species.
Taxonomy (*tax* = to arrange) The science of organizing and classifying organisms.
Teleosts A large category of bony fish.
Terrigenous Describing sediments that originate from land-based sources.
Test The **calcareous** endoskeleton of **echinoids.** In sea urchins, it is globe-shaped, while in sand dollars, it is disk-shaped.
Thallus The entire body of a seaweed.
Thermocline The depth where water temperature begins to drop abruptly below the surface temperature.
Thorax The middle body region or arthropods, often where legs are attached.
Tide The periodic change in sea level due to the gravitational influence of the sun and moon upon the ocean surface.
Tide pool A place in the rocky **intertidal zone** where seawater is retained after the tide recedes.
Tide table A printed table of tidal heights and the times they occur, for specific coastal locations.
Tintinnids Unicellular aquatic **protists** with a shell-like covering called a **lorica.**
Topography The study of the terrain and three-dimensional surface features of the earth.
Transect A study where data is collected along a continuous path. To conduct a beach transect to show changes in elevation, transects are conducted perpendicular to the shore.
Trochophore The ciliated planktonic larval form of annelids and some mollusks.
Trophic level The position of an organism in a **food chain,** showing whether it is a **primary producer, primary** or **secondary consumer**, or **decomposer**.
Tube feet Flexible, extensible organs on the surface of echinoderms, used for locomotion and food capture. Part of the **water vascular system**.
Tunic The leathery or rubbery outer casing of **tunicates**.
Tunicates See **Sea Squirts**.

Turbellarians Free-living members of the phylum Platyhelminthes (flatworms).

Tusk shells Members of the phylum Mollusca, class Scaphopoda. Benthic mollusks that inhabit a pointed, tube-like shell.

Typhoon A hurricane that occurs in the northwest Pacific Ocean, west of the International Dateline.

Umbo Synonymous with the **beak** of bivalves.

Upwelling The vertical movement of deep water up to or near the ocean surface; **phytoplankton blooms** occur in areas of upwelling, due to the higher levels of nutrients brought to the surface.

Urban runoff Rainfall that runs off city streets into storm drains and receiving waters (streams, lakes, the ocean, etc); usually delivering many types of pollutants and chemicals into those waters.

Urogenital opening An opening where excretory waste and gametes leave the body. On fish, it is **posterior** to the anus. On marine mammals, it is **anterior** to the anus.

Valve 1) One of the **calcareous** halves of a clam, mussel, or other **bivalve** shell; 2) One of the **siliceous** halves of a diatom **frustule**.

Veliger The ciliated, planktonic larval stage of mollusks.

Ventral The opposite of **dorsal**.

Vertebrates Members of the subphylum Vertebrata; animals that have a cranium and a segmented spinal column composed of repeating vertebrae.

Vivipary (*viv* = alive) Reproduction when fertilization is internal and the young are nourished through a physical connection to the mother's bloodstream (*Adj.* **viviparous**). A characteristic of mammals and some sharks.

Warning coloration Obvious markings on an animal's body to serve as a warning to predators that the animal has toxins or distasteful chemicals as its defense.

Water-vascular system A unique anatomical feature of echinoderms consisting of a complex internal system of connected tubes running throughout the echinoderm body. Water is pumped through the various canals to operate the **tube feet**.

Whalebone See **Baleen**.

Whorl One complete coil on the spiral shell of a univalve gastropod mollusk.

Windward The side of a reef or island that faces the predominant wind direction.

Wrack line A line of debris cast ashore onto the beach during a previous high tide. It can contain seaweed, trash, driftwood, and other floating materials. Also called the **strand line**.

Xanthophylls (*xanth* = yellow, *phyl* = leaf) A group of golden-yellowish **accessory pigments** in some types of algae.

Zoea A planktonic larval stage of crabs; the developmental stage preceding the **megalops** larva.

Zonation A characteristic pattern of horizontal bands formed by species distributions in the rocky intertidal zone.

Zooid An individual member of a clonal colony; however, each zooid remains connected to the colony and cannot survive on its own.

Zooplankton Animal members of the planktonic community.

Zooxanthellae Symbiotic **dinoflagellates** living within the tissues of reef-building corals and certain sea anemones. The photosynthetic dinoflagellates produce substances used by the host animal, and the host provides both nutrients and shelter to the zooxanthellae. Thus, this relationship is an example of **mutualism**.

Zygote (*zyg* = a yoke) A **diploid** fertilized egg, formed from the fusion of **haploid gametes**.

APPENDIX C

Bibliography of Field & Identification Guides for North American Marine Organisms

PACIFIC COAST GUIDES:

Abbott, R. T. 2001. *Seashells of North America: A Guide to Field Identification.* New York, NY: St. Martin's Press.

Abbott, R. T. and P. A. Morris. 2001. *A Field Guide to Shells: Atlantic and Gulf Coasts and the West Indies (Peterson Field Guides).* New York, NY: Houghton Mifflin.

Allen, G. R. and D. R. Robertson. 1994. *Fishes of the Tropical Eastern Pacific.* Honolulu, HI: University of Hawaii Press.

Brusca, G. J. and R. C. Brusca. 1979. *A Naturalist's Seashore Guide: Common Marine Life of the Northern California Coast and Adjacent Shores.* Eureka, CA: Mad River Press.

California Department of Fish and Game. 1972. *Guide to the Coastal Marine Fishes of California.* California Fish Bulletin Number 157. Division of Agricultural Sciences: University of California.

Carlton, J. T. (ed.) 2007. *The Light and Smith Manual: Intertidal Invertebrates from Central California to Oregon.* Berkeley, CA: University of California Press.

Dawson, E. Y. and M. S. Foster. 1982. *Seashore Plants of California.* Berkeley, CA: University of California Press.

Denny, M. W. and S. Gaines (eds.) 2007. *Encyclopedia of Tidepools and Rocky Shores (Encyclopedias of the Natural World).* Berkeley, CA: University of California Press.

Druehl, L. 2001. *Pacific Seaweeds.* Madeira Park, British Columbia: Harbour Publishing.

Eder, T. and I. Sheldon. 2001. *Whales and Other Marine Mammals of Oregon and Washington.* Auburn, WA: Lone Pine Press.

Eder, T. and I. Sheldon. 2002. *Whales and Other Marine Mammals of California and Baja.* Auburn, WA: Lone Pine Press.

Eschmeyer, W. N. and E. S. Herald. 1983. *A Field Guide to Pacific Coast Fishes: North America: From the Gulf of Alaska to Baja California (Peterson Field Guides).* Wilmington, MA: Houghton Mifflin.

Fichter, G. S. and B. J. Hoopes. 1990. *Whales, Dolphins, and Other Marine Mammals.* New York, NY: St. Martin's Press.

Gotshall, D. W. 1998. *Sea of Cortez Marine Animals.* Gig Harbor, WA: Sea Challengers Press.

Gotshall, D. W. 2001. *Pacific Coast Inshore Fishes.* Gig Harbor, WA: Sea Challengers Press.

Harbo, R. 1999. *Whelks to Whales: Coastal Marine Life in the Pacific Northwest.* Madeira Park, British Columbia: Harbour Publishing.

Higgins, R. P. and Thiel, H. 1988. *Introduction to the Study of Meiofauna.* Washington, DC: Smithsonian Institution Press.

Hinton, S. 1987. *Seashore Life of Southern California.* Berkeley, CA: University of California Press.

Humann, P. 1996. *Coastal Fish Identification: California to Alaska.* Jacksonville, FL: New World Publications.

Jensen, G. C. 1995. *Pacific Coast Crabs and Shrimps.* Gig Harbor, WA: Sea Challengers Press.

Kozloff, E. N. 1983. *Seashore Life of the Northern Pacific Coast.* Seattle, WA: University of Washington Press.

Kozloff, E. N. 1996. *Marine Invertebrates of the Pacific Northwest.* Seattle, WA: University of Washington Press.

Lamb, A. and B. P. Hanby. 2005. *Marine Life of the Pacific Northwest: A Photographic Encyclopedia of Invertebrates, Seaweeds, and Selected Fishes.* Madeira Park, British Columbia: Harbour Publishing.

Leatherwood, S. and R. R. Reeves. 1983. *The Sierra Club Handbook of Whales and Dolphins.* San Francisco, CA: Sierra Club Books.

Mondragon, J. and J. Mondragan. 2003. *Seaweeds of the Pacific Coast: Common Marine Algae from Alaska to Baja California.* Gig Harbor, WA: Sea Challengers Press.

Morris, H. R., and E. C. Haderlie. 1980. *Intertidal Invertebrates of California.* Stanford, CA: Stanford University Press.

O'Clair, R. M. and S. C. Lindstrom. 2000. *North Pacific Seaweeds.* Auke Bay, AK: Plant Press.

Orr, R. T., Helm, R., and J. Schoehwald. 1989. *Marine Mammals of California.* Berkeley, CA: University of California Press.

Ricketts, E. F., Calvin, J., Hedgpeth, J. W., and D. W. Phillips. 1985. *Between Pacific Tides* (5th ed.). Stanford, CA: Stanford University Press.

Smith, D. L. and K. B. Johnson. 2003. *A Guide to Marine Coastal Plankton and Marine Invertebrate Larvae.* Dubuque, IA: Kendall/Hunt Publishing Company.

Waaland, J. R. 1977. *Common Seaweeds of the Pacific Coast.* Seattle, WA: Pacific Search Press.

GULF COAST GUIDES:

Abbott, R. T. 2001. *Seashells of North America: A Guide to Field Identification.* New York, NY: St. Martin's Press.

Abbott, R. T. and P. A. Morris. 2001. *A Field Guide to Shells: Atlantic and Gulf Coasts and the West Indies (Peterson Field Guides).* New York, NY: Houghton Mifflin.

Fichter, G. S. and B. J. Hoopes. 1990. *Whales, Dolphins, and Other Marine Mammals.* New York, NY: St. Martin's Press.

Goodson, G. 1976. *Fishes of the Atlantic Coast: Canada to Brazil, Including the Gulf of Mexico, Florida, Bermuda, the Bahamas, and the Caribbean.* Stanford, CA: Stanford University Press.

Higgins, R. P. and Thiel, H. 1988. *Introduction to the Study of Meiofauna.* Washington, DC: Smithsonian Institution Press.

Hoese, H. D., Moore, R. H., and A. F. Calhoun. 1998. *Fishes of the Gulf of Mexico: Texas, Louisiana, and Adjacent Waters (W.L. Moody, Jr., Natural History Series, No 22).* College Station, TX: Texas A & M University Press.

Humann, P. and N. DeLoach. 1995. *Snorkeling Guide to Marine Life: Florida, Caribbean, Bahamas.* Jacksonville, FL: New World Publications.

Humann, P. and N. DeLoach. 2002. *Reef Fish Identification: Florida, Caribbean, Bahamas.* Jacksonville, FL: New World Publications.

Johnson, W. S., Allen, D. M., and M. Fyling. 2005. *Zooplankton of the Atlantic and Gulf Coasts: A Guide to Their Identification and Ecology.* Baltimore, MD: Johns Hopkins University Press.

Kaplan, E. H. 1999. *A Field Guide to Southeastern and Caribbean Seashores: Cape Hatteras to the Gulf Coast, Florida, and the Caribbean.* Boston, MA: Houghton-Mifflin.

Leatherwood, S. and R. R. Reeves. 1983. *The Sierra Club Handbook of Whales and Dolphins.* San Francisco, CA: Sierra Club Books.

Littler, D. A., Littler, M.M., Bucher, K.E., and J. N. Norris. 1989. *Marine Plants of the Caribbean: A Field Guide from Florida to Brazil.* Washington, DC: Smithsonian Institution Press.

Schneider, C. W. and R. B. Searles. 1991. *Seaweeds of the Southeastern United States.* Durham, SC: Duke University Press.

Smith, C. L. 1997. *National Audobon Society Field Guide to Tropical Marine Fishes of the Caribbean, Gulf of Mexico, Florida, the Bahamas, and Bermuda.* New York, NY: Knopf.

Smith, D. L. and K. B. Johnson. 2003. *A Guide to Marine Coastal Plankton and Marine Invertebrate Larvae.* Dubuque, IA: Kendall/Hunt Publishing Company.

Voss, G. L. 1976. *Seashore Life of Florida and the Caribbean.* Miami, FL: Banyan Books.

EAST COAST GUIDES:

Abbott, R. T. 2001. *Seashells of North America: A Guide to Field Identification.* New York, NY: St. Martin's Press.

Abbott, R. T. and P. A. Morris. 2001. *A Field Guide to Shells: Atlantic and Gulf Coasts and the West Indies (Peterson Field Guides).* New York, NY: Houghton Mifflin.

Carpenter, K. E. 2002. *Living Marine Resources of the Western Central Atlantic, Vol. 1 (FAO Species Identification Guide for Fishery Purposes).* FAO ISSN 1020-6868 ISBN 92-5-104825-8

Clark, A. M. and M. E. Downey. 1991. *Starfishes of the Atlantic: An Illustrated Key.* New York, NY: Chapman and Hall.

Denny, M. W. and S. Gaines (eds.) 2007. *Encyclopedia of Tidepools and Rocky Shores (Encyclopedias of the Natural World).* Berkeley, CA: University of California Press.

Fichter, G. S. and B. J. Hoopes. 1990. *Whales, Dolphins, and Other Marine Mammals.* New York, NY: St. Martin's Press.

Gerber, R.P. 2000. *An Identification Manual to the Coastal and Estuarine Zooplankton of the Gulf of Maine Region.* Brunswick, ME: Acadia Productions.

Goodson, G. 1976. *Fishes of the Atlantic Coast: Canada to Brazil, Including the Gulf of Mexico, Florida, Bermuda, the Bahamas, and the Caribbean.* Stanford, CA: Stanford University Press.

Gosner, K. L. and R. T. Peterson. 1999. *A Field Guide to the Atlantic Seashore: From the Bay of Fundy to Cape Hatteras.* New York, NY: Houghton-Mifflin.

Higgins, R. P. and Thiel, H. 1988. *Introduction to the Study of Meiofauna.* Washington, DC: Smithsonian Institution Press,

Hillson, C. J. 1997. *Seaweeds: A Color-coded, Illustrated Guide to Common Marine Plants of the East Coast of the United States.* University Park, PA: Pennsylvania State University Press.

Johnson, W. S., Allen, D. M., and M. Fyling. 2005. *Zooplankton of the Atlantic and Gulf Coasts: A Guide to Their Identification and Ecology.* Baltimore, MD: Johns Hopkins University Press.

Katona, S., Richardson, D., and V. Rough. 1993. *A Field Guide to the Whales, Porpoises, and Seals from Cape Cod to Newfoundland.* Washington, DC: Smithsonian Institution Press.

Leatherwood, S. and R. R. Reeves. 1983. *The Sierra Club Handbook of Whales and Dolphins.* San Francisco, CA: Sierra Club Books.

Lee, T. F. 1977. *The Seaweed Handbook: An Illustrated Guide to Seaweeds from North Carolina to the Arctic.* New York, NY: Dover Publications.

Lippson, A. J. and R. L. Lippson. 1997. *Life in the Chesapeake Bay.* Baltimore, MD: Johns Hopkins University Press.

Martinez, A. 2003. *Marine Life of the North Atlantic, 3rd Edition: Canada to New England.* Locust Valley, NY: Aqua Quest Publications.

Murdy, E. O., Birdsong, R. S., and J. A. Musick. 1996. *Fishes of Chesapeake Bay.* Washington, DC: Smithsonian Institution Press.

Pollack, L. W. 1998. *A Practical Guide to the Marine Animals of Northeastern North America.* New Brunswick, NJ: Rutgers University Press.

Ray, G. C., Robins, C. R., Douglass, J., and R. T. Peterson. 1999. *A Field Guide to Atlantic Coast Fishes: North America (Peterson Field Guides).* New York, NY: Houghton Mifflin.

Ruppert, E. and R. Fox. 1988. *Seashore Animals of the Southeast.* Columbia, SC: University of South Carolina Press.

Villalard-Bohnsack, M. 1995. *Illustrated Key to the Seaweeds of New England.* Kingston, RI: The Rhode Island Natural History Survey.

Watling, L., Fegley, J., and J. Moring. 2003. *Life Between the Tides: Marine Plants and Animals of the Northeast.* Gardiner, ME: Tilbury House Publishers.

Weiss, H. M. 1995. *Marine Animals of Southern New England and New York.* Connecticut: State Geological and Natural History Survey of Connecticut, Department of Environmental Protection.

APPENDIX D

Depth Zonation of the World Ocean

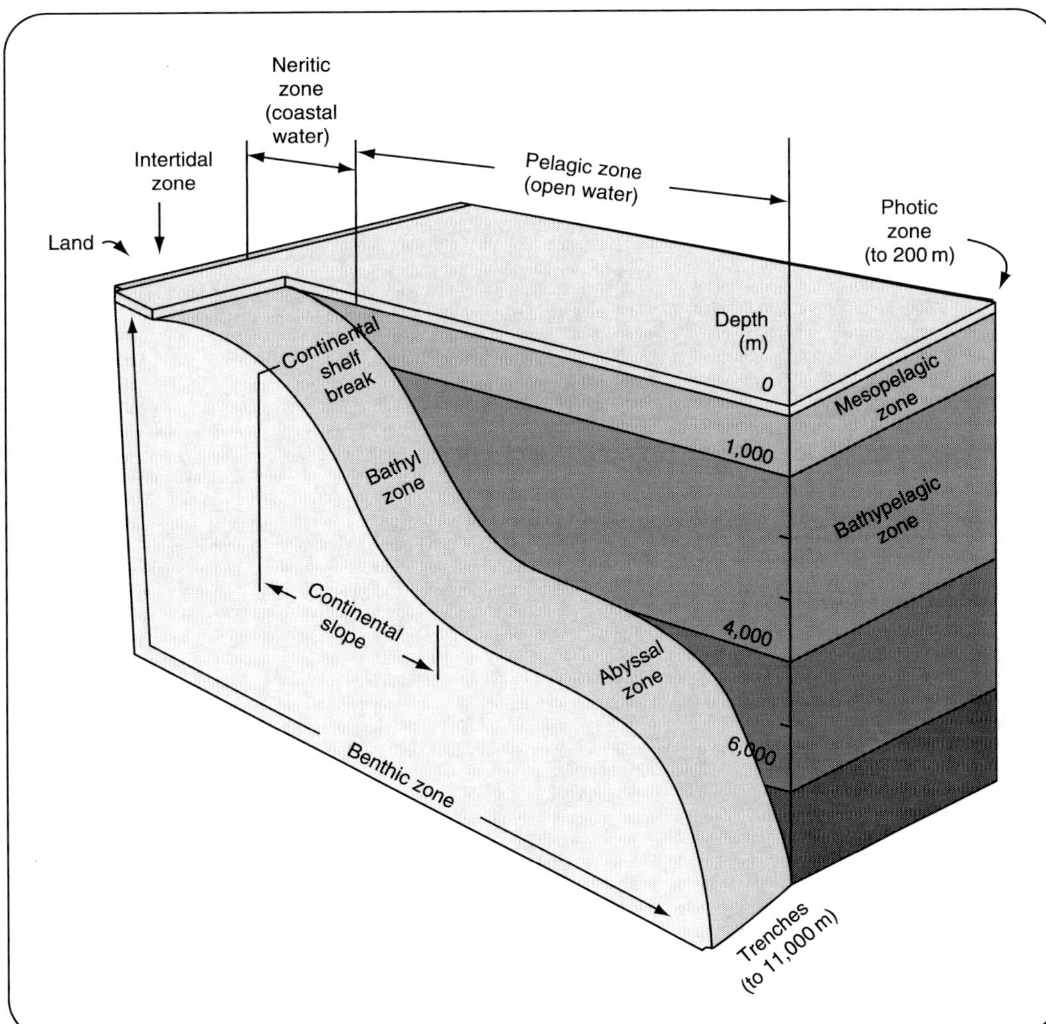

Oceans and the Seas

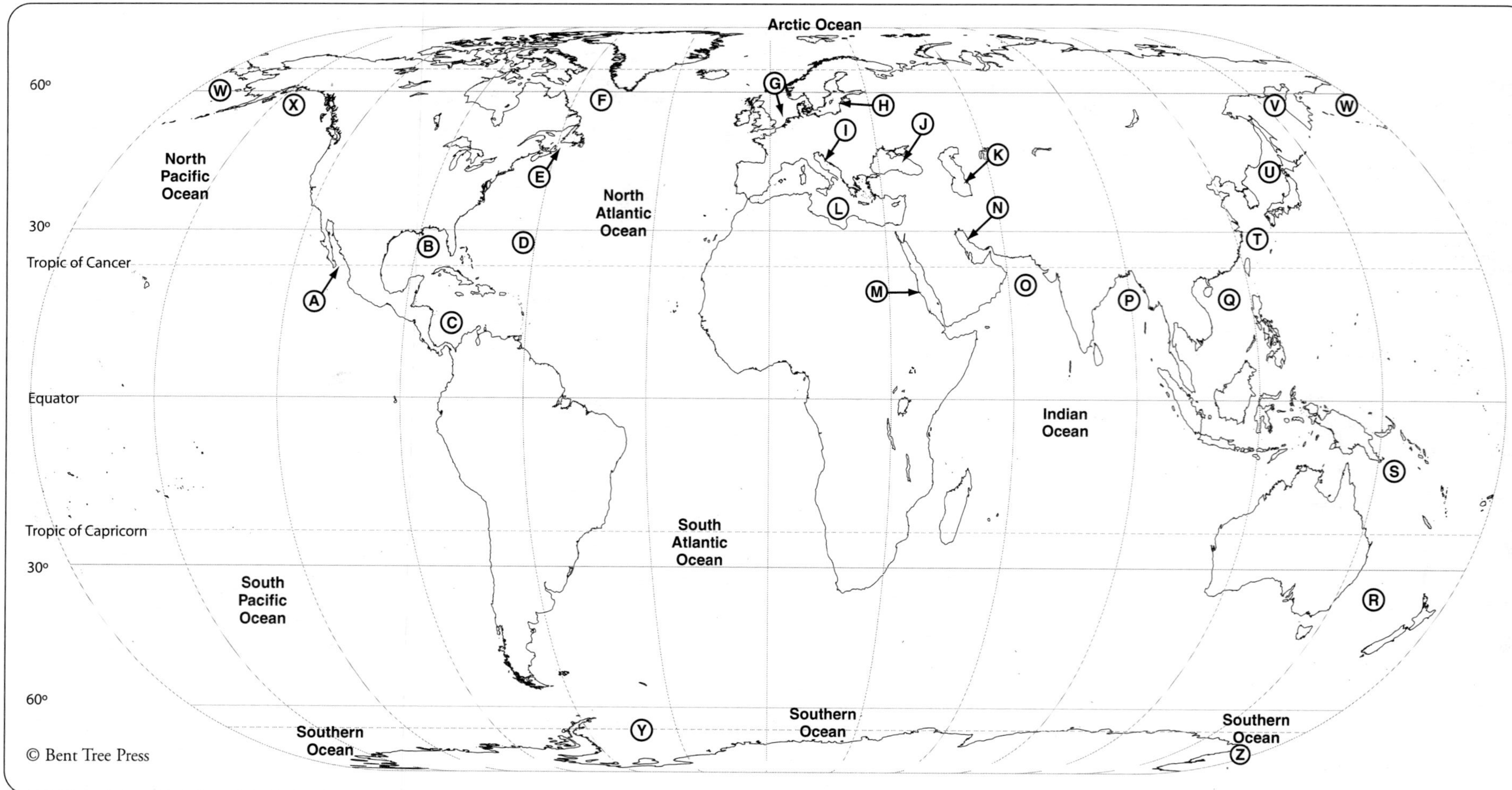

A – Gulf of California
B – Gulf of Mexico
C – Caribbean Sea
D – Sargasso Sea
E – Gulf of St. Lawrence
F – Labrador Sea
G – North Sea
H – Baltic Sea
I – Adriatic Sea
J – Black Sea
K – Caspian Sea
L – Mediterranean Sea
M – Red Sea
N – Persian Gulf
O – Arabian Sea
P – Bay of Bengal
Q – South China Sea
R – Tasman Sea
S – Coral Sea
T – Yellow Sea
U – Sea of Japan
V – Sea of Okhotsk
W – Bering Sea
X – Gulf of Alaska
Y – Weddell Sea
Z – Ross Sea